# AN OCEAN OF AIR

## A Natural History of the Atmosphere

Gabrielle Walker

BLOOMSBURY

First published in Great Britain 2007



Diagrams by John Gilkes



Bloomsbury Publishing Plc
36 Soho Square
London W1D 3QY

www.bloomsbury.com

Bloomsbury Publishing, London, New York and Berlin

A CIP catalogue record for this book is available from the British Library

ISBN 9780747581901

10 9 8 7 6 5 4 3 2 1

Typeset by Hewer Text UK Ltd, Edinburgh
Printed in Great Britain by Clays Ltd, St Ives Plc

The paper this book is printed on is certified by the © 1996 Forest Stewardship Council A.C. (FSC). It is ancient-forest friendly. The printer holds FSC chain of custody SGS-COC-2061

For Fred and Hubert

*I adorn all the earth.*
*I am the breeze that nurtures all things green.*
*I encourage blossoms to flourish with ripening fruits.*
*I am led by the spirit to feed the purest streams.*
*I am the rain coming from the dew*
*That causes the grasses to laugh with the joy of life.*

– Hildegard of Bingen
Twelfth-century abbess

# CONTENTS

# PROLOGUE

**16 August 1960, 7 A.M.**

Thirty-two kilometres above New Mexico, Joe Kittinger was hanging in the sky. For eleven minutes he remained there, poised in an open gondola that twirled slowly beneath a vast helium balloon. Though it was long past sunrise, the air around was dark as midnight. Far below, where Earth's surface curved away to the horizon, a glowing blue halo stood out against the blackness of space.

This glow was the atmosphere, the single greatest gift our planet possesses. Earth's glorious blue colour comes not from the oceans, but from the sky, and every astronaut who has seen that delicate halo has come back with the same tale: they couldn't believe how fragile it made Earth seem, and how beautiful.

Back on the surface, robbed of that lofty perspective, we are inclined to take our atmosphere for granted. Yet air is one of the most miraculous substances in the universe. Single-handedly, that thin blue line has transformed our planet from a barren lump of rock into a world full of life. And it is the only shield that stands between vulnerable earthlings and the deadly environment of space.

Kittinger, however, had journeyed beyond its protection. Up on the edge of space, the air was so tenuous that if his pressure suit failed he would die within minutes. First his saliva would bubble, then his eyes pop and his stomach swell, and finally his blood would boil. Despite all the risks he had taken as a test pilot for the US Air Force, he had never been in greater danger.

Alone in his gondola, he was acutely aware of the menace. The near-vacuum seemed strangely substantial, like an enveloping layer of poison. The darkness rattled him, as did the curtain of clouds far below that were cutting off all views of home. He radioed to ground control. 'There is a hostile sky above me,' he said. 'Man will never conquer space. He may live in it, but he will never conquer it.'

He shuffled towards the door, weighed down with 150 pounds of survival gear, instruments and cameras, and stood for a moment, his boots protruding slightly over the ledge. Several inches below his feet, a sign declared 'highest step in the world'. He took a single breath of pure oxygen from within his tightly sealed helmet. 'Lord, take care of me now,' he said. And then he jumped.

At first, Kittinger had no sense of falling. He could see the white swirls of storm clouds far below his feet, but they were growing no closer. The air around him was so thin that there was no sound or wind or any other clue that he was plunging through the most hostile environment a human being had ever faced. Spread-eagled in the sky, he felt almost serene. He could have been floating on a sea of nothingness.

Dangerous though the environment was, it was still protecting him. Lack of pressure isn't the only hazard in space; there is also a constant barrage of radiation, much of it from our own sun. Every day, along with the warmth and light that allow us to live on Earth, the sun sends out x-rays and ultraviolet light from the deadly end of its rainbow.

Thanks to our intervening sky, this radiation never reaches the ground. Eighty kilometres above Kittinger's head, a few scarce atoms of air were acting as sentinels, intercepting and absorbing those lethal x-rays. In the process, these atoms were being ripped to shreds and heated to temperatures of 2,000 degrees. They form the ionosphere, a tenuous region of the atmosphere where electricity is king. Unseen from Earth's surface, giant blue jets of fire leap up to this layer from the tops of thunderclouds, in bolts of reverse lightning. Here, meteorites from space are obliterated in the glorious blazes of light that we call shooting stars. They splatter the air with floating layers of metal that allow electric currents to swoop around the upper Earth. Radio broadcasts reflect from this charged surface as they bounce their way around the globe.

Farther up still, the air above Kittinger was facing an even more violent attack – from a force known as the solar wind. Electrically charged jets of particles from the sun were barrelling towards Earth at more than a million kilometres an hour, ready to strip away our atmosphere and send it streaming out behind the planet like the tail from a giant comet.

But to do so, it would first need to pass one of our staunchest defenders – Earth's magnetic field. On the surface we scarcely notice this field, except when it helpfully tugs compass needles into pointing north. But its arching influence extends tens of thousands of kilometres above us, and it forces the solar wind to part around it like water around a ship's bow. Far above Kittinger's head, those protective magnetic arcs were channelling the solar wind harmlessly away. The field is all but impenetrable, allowing just a few particles to leak into the polar regions, where they collide with the atmosphere to provide the dancing glows of the northern and southern lights.

Still, almost all our protective atmosphere lies within a few miles of the surface, and when Kittinger took that high-altitude leap of faith, most of it was beneath him. A few seconds into his fall, he kicked and twisted until he was facing upwards. Now he could see the taut white sphere of his balloon shooting up into the darkness at a breakneck speed. This, Kittinger knew, was an illusion. The balloon was still floating gently where he had left it. *He* was the one falling down through the sky at close to the speed of sound.

Kittinger was tumbling now through another of our world's vital protective shields – the ozone layer. All around him, any invisible ultraviolet rays that had slipped through the ionosphere were being soaked up by a diffuse cloud of invisible gas. Ozone is miraculous stuff. Near the ground it is sometimes created by lightning bolts or spark plugs. It smells like burning electrical wires and makes you choke. But high aloft it is both vigilant and resilient. Split asunder by ultraviolet rays, the ozone molecules around Kittinger were calmly re-forming. Like the burning bush encountered by Moses, they are constantly ablaze but never consumed.

Twenty thousand metres. Eighteen thousand. Kittinger was now below the point where even a pinhole in his suit would have allowed his blood to boil off into space. But he had one last hazard to face: he had reached the coldest part of his descent, where the temperature had fallen to –72 degrees Celsius and the heating elements in his suit mattered most.

Then there were clouds, and wind, and all the signs that Kittinger was finally approaching home. Twelve thousand metres. Ten thousand. He was about to drop past the altitude of Mount Everest. Any jet plane that happened to be flying nearby would see a man in a strange suit shooting past the window. The clouds he had seen from the gondola, blocking his view of home, were now rushing towards him. Though he knew they were only insubstantial water droplets, he still braced himself unconsciously for an impact, pulling his legs upward in anticipation. The moment he hit the clouds, his parachute opened and he knew he would live. 'Four minutes and thirty-seven seconds free fall!' he said into his voice recorder. 'Ahhhhh boy!'

Kittinger had now fallen safely into the lowest part of the atmosphere – the troposphere. The air here isn't so much a protector as a transformer, a thick, life-giving blanket of air, wind and weather that turns our planet into home. After the bone-dryness of space, flecks of moisture from the clouds fogged

Kittinger's face plate. He could feel the tug of the thickening air. The sky was now full of life, though he couldn't see it. Bacteria that had launched themselves into the wind were hitching a ride on cloud droplets, seeking out new victims farther afield. Insects were wafting their way to new feeding grounds and seeds to more fertile soil.

And, praise be, two rescue helicopters were hovering nearby. With the ground fast approaching, Kittinger struggled to cut away his heavy instrument kit for the sake of a softer landing, but one last hose resisted his knife. He gave up and instead raised the shield on his helmet and took a deep breath of fresh air. As the air flushed into his lungs, oxygen leapt across thin membranes into the cells of his blood and turned them a glorious, life-giving red. (And some of it set off on a well-worn rampage that had been going on since Kittinger took his first breath of air. These rogue molecules would continue to line Kittinger's face and wear down his body, in the process we know as ageing.)

Finally, after a flight time of thirteen minutes and forty-five seconds, Joe Kittinger crashed unceremoniously into the scrub, twenty-seven miles west of Tularosa, New Mexico. Medical personnel, ground crew, supporters and journalists poured out of the helicopters and rushed over to where he lay. He smiled at them through his open face plate. 'I'm very glad to be back with you all,' he said. Though the desert landscape was far from lush, to a man who had seen beyond the atmosphere, the yuccas and sagebrush seemed full of life. 'Fifteen minutes before I'd been on the edge of space,' he said later. 'And now, to me, I was in the Garden of Eden.'

Captain Joseph W. Kittinger Jr. of the US Air Force is the man who fell to Earth and lived. Nobody has ever managed to emulate his feat. His passage home from the edge of space, from thin air to thick, illustrates something extraordinary about our planet. Space is almost close enough to touch. Only thirty-two kilometres above our heads is an appalling, hostile environment that would freeze us, and burn us and boil us away. And yet our enfolding layers of air protect us so completely that we don't even realise the dangers. This is the message from Kittinger's flight, and from every one of the pioneers who have sought to understand our atmosphere: we don't just live *in* the air. We live *because of* it.

# PART ONE

# COMFORT BLANKET

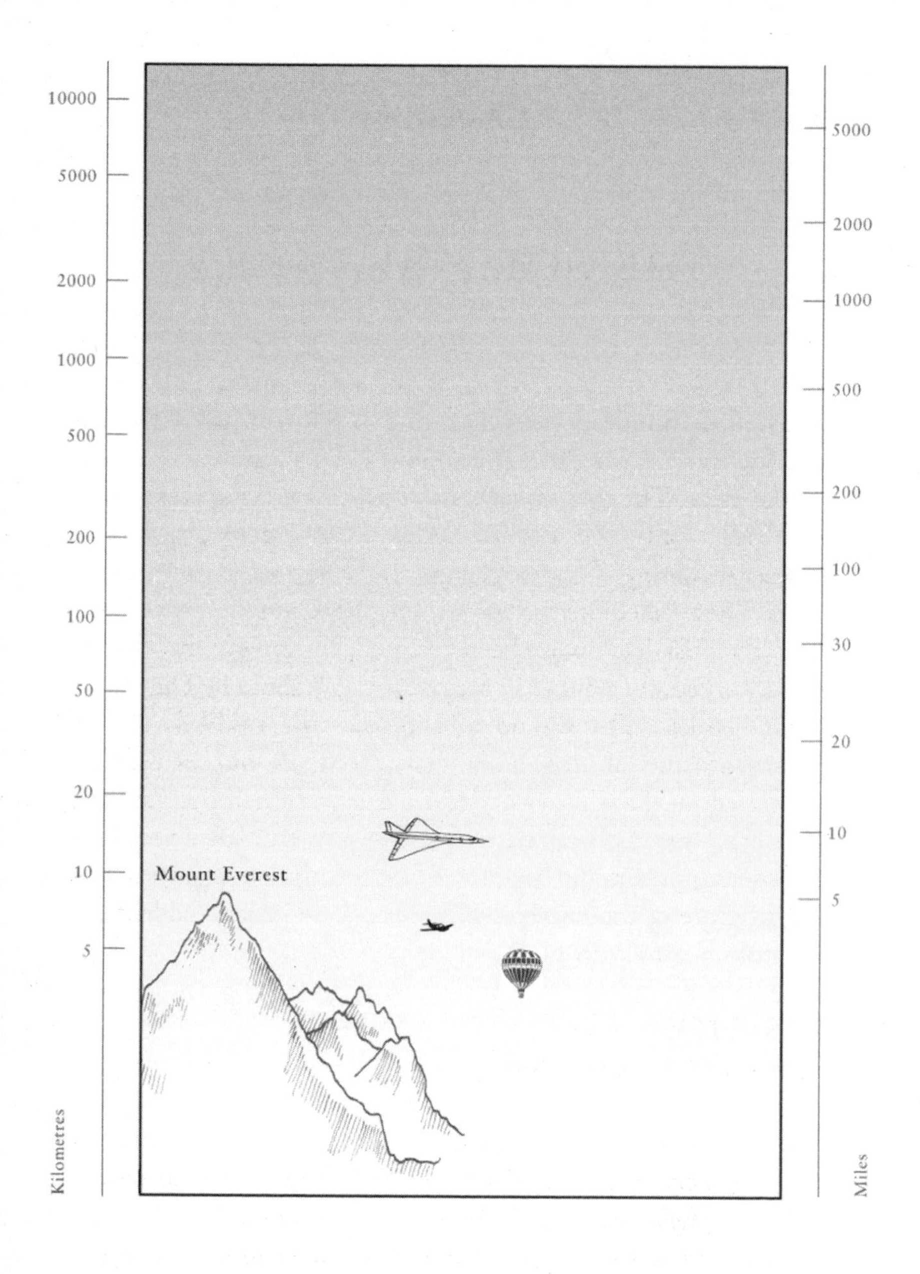
10000
5000
2000
1000
500
200
100
50
20
10
5
Kilometres
Mount Everest
5000
2000
1000
500
200
100
30
20
10
5
Miles

# *Chapter 1*

# THE OCEAN ABOVE US

Nearly four hundred years ago, in a patchwork of individual fiefdoms that we now call Italy, a revolution of ideas was struggling to take place. The traditional way to understand the workings of the world – through a combination of divine revelation and abstract reasoning[1] – had begun to come under attack from a new breed. These called themselves 'natural philosophers', because the word 'scientist' had not yet been invented. To find out the way the world worked, they didn't sit around and talk about it. They went out and looked. This was not an approach that was likely to find favour with the Church, home of received wisdom, or with its instruments – the whispering Inquisitors, with their hotline back to Rome. Now, a certain natural philosopher had fallen very foul of those Inquisitors and been forced to stop his investigations into the structure of the heavens. His name was Galileo Galilei, and our story begins with him.

**22 June 1633**
**Convent of Minerva, Rome**

> *I, Galileo Galilei, son of the late Vincenzo Galilei, Florentine, aged seventy years, arraigned personally before this tribunal, and kneeling before you, most Eminent and Reverend Lord Cardinals, Inquisitors general against heretical depravity throughout the whole Christian Republic . . . have been pronounced by the*

*Holy Office to be vehemently suspected of heresy, that is to say, of having held and believed that the sun is the centre of the world and immovable, and that the earth is not the centre and moves:*

*Therefore, desiring to remove from the minds of your Eminences, and of all faithful Christians, this strong suspicion, reasonably conceived against me, with sincere heart and unfeigned faith I abjure, curse, and detest the aforesaid errors and heresies . . . and I swear that in the future I will never again say or assert, verbally or in writing, anything that might furnish occasion for a similar suspicion regarding me.*

As the great Galileo rose from his knees at the end of this infamous, and forced, recantation, he is said to have muttered, '*Eppur si muove!*' ('And yet it moves!'). He knew in his heart that the Earth moves around the sun, in spite of what the Inquisitors had made him say. Still, devoutly religious as he was, he had no taste for defying his own church. Nor had he any desire to share the fate of the unfortunate monk Giordano Bruno, who a few decades earlier had been publicly burned for holding similar views. Galileo may have been the most famous philosopher in all Italy, but he knew that in itself wouldn't save him from the fire.

And though he was now seventy years old, frail and steadily losing his sight, he was not yet ready to die. He had damaged his eyes by staring through a telescope at wonders he himself had discovered: blemishes that appeared periodically on the surface of the sun; craters on the moon; distant but distinct moons circling the planet Jupiter (who would have thought that other planets could have moons of their own?), and stars that nobody knew existed. Now, before the cataracts and glaucoma finally clouded his sight, in secret if necessary, he had one last task to complete. Galileo had seen this 'trial' coming; he'd known for some time that he couldn't continue his study of the heavens. So for some years now he had been discreetly changing tack, turning his attention inwards to Earth itself. And, failing eyesight notwithstanding, he was about to change the way we see the most apparently ordinary substance in the world: air.

The Inquisitors knew nothing of this. They were satisfied with his recantation, and decided, graciously, to spare his life. He would be allowed to return to his villa at Arcetri in Florence, though he should understand that he was still considered dangerous and would therefore be held under house arrest. There would be no visitors, save those given prior permission by the Church. Meanwhile, Galileo himself was to spend his time reciting the holy psalms as penance, and praying for his immortal soul.

Galileo returned to his villa as instructed and performed his penance diligently. But the Inquisitors had also obliged him to swear never again to publish work that might offend the Holy Office, and he had no intention of complying. For with him to Arcetri he had taken a certain manuscript that was already nearly finished.

He had started the experiments it described while awaiting his summons to Rome. Having turned away from his telescope, Galileo had become fascinated instead by the different ways that objects move through the air.[2] The result was to become his masterpiece. The manuscript already recounted findings that would become just as famous as the moons of Jupiter. For instance, Galileo had made the surprising discovery that Earth's gravity doesn't care in the least how much something weighs. Drop a cannon-ball and a pebble from a high tower, and both will reach the ground at exactly the same moment.[3]

But within its pages was another discovery that would prove to be less famous yet no less significant. Galileo had measured the weight of air.

This might seem like a bizarre notion. How can something so insubstantial as the air weigh anything at all? In fact our planet's air is constantly pushing down on us with great force. We don't notice this because we're used to it, like lobsters sauntering along on the seafloor, unaware of the crushing weight of the ocean of water above them. We give our own overlying air-ocean so little respect that we even describe anything that's full of air as being 'empty'.

Back in Galileo's time, notions about air were similarly hazy. Most people accepted the idea put forward by Aristotle in the fourth century BC that everything in the world was made up of four elements: earth, air, fire and water. Earth and water were obviously pulled downwards by gravity. Fire was obviously weightless. But air was the problem child. Was it heavy enough to be dragged to the ground, light enough to rise like flames do, or did it simply ignore Earth's gravitational tug and hover?

Galileo believed that air is heavy and had set about testing his idea. The experiments he performed were typically ingenious. First, he took a large glass bottle with a narrow neck and a tight leather stopper. Into this stopper he inserted a syringe attached to a bellows and by working vigorously managed to squeeze two or three times more air into the bottle than it had previously contained. Next, he weighed the glass bottle most precisely, adding and subtracting the finest of sand to his scales until he was satisfied with the answer. Then, he opened a valve in the lid. Immediately, the compressed air rushed out of its confinement, and the bottle was suddenly a handful of grains lighter. The air that had escaped must account for the missing weight.

This showed that air is not the insubstantial body we usually take it for. But now Galileo wanted to know how much air corresponded to how many grains of sand. For that he would somehow need to measure both the weight of the escaping air and its volume.

This time, he took the same glass bottle with its long narrow neck. However, instead of pumping it full of extra air, he forced in some water. When the bottle was three-quarters full of water, its original air was squeezed uncomfortably into a quarter of its original space. Galileo weighed the bottle accurately, opened the valve, allowed this pressurised air to escape, and then weighed the bottle again to find out how much air he had lost. As for the volume, Galileo reasoned that the portion of air that had been forced to leave the bottle had been pushed aside by the water he had squeezed in. So the volume of air that had fled must be exactly the same as the volume of water that remained. All he had to do was pour out the water and measure its volume and *voilà*, he had found the weight for a given volume of air.

The value Galileo came up with was surprisingly large: air seemed to weigh as much as one four-hundredth the weight of an equivalent amount of water.[4] If that doesn't sound like much, consider this. Picture a particular volume of air for a moment – such as the 'empty' space inside the Albert Hall in London. How heavy would you expect that amount of air to be? Would it weigh ten kilos? Or a hundred? Or maybe even five hundred?

The answer is somewhere in the region of thirty tonnes.

The weight of air is so extreme that even Galileo didn't see the whole story. He never considered the question of how we can shoulder such a crushing, overwhelming burden, for the simple reason that he didn't realise the air *above us* is still heavy. He had measured the weight of air in his bottle, but he was convinced that the moment this air was released back into its natural element, the sky, it immediately ceased to weigh anything at all.[5] Galileo believed that our atmosphere as a whole is incapable of pushing. It was one of the few occasions when the great man was wrong.

In spite of the Church's opposition, Galileo finished his manuscript – and published it. After fruitless efforts to convince publishers in Florence, Rome and Venice to defy the Inquisitors, Galileo finally smuggled the manuscript out to a printer in the Netherlands. Four years later, as he approached the end of his life, a few copies began filtering back to Italy. Each bore a disingenuous disclaimer by Galileo himself, who wrote how astonished he was that his words had somehow found their way to a printer's in spite of his obedience to the papal diktat.

And although Galileo was wrong about the way our air

behaves aloft, the experiments his great work contained would influence two very different people to discover the truth.

By coincidence, both of these people arrived in Florence at more or less the same time, in October 1641, just a few months before Galileo's death. One was a 33-year-old Roman mathematician named Evangelista Torricelli, who had been working with Galileo in the final three months before his death.

Torricelli had become fascinated by Galileo's experiments on air and his conviction that, though air was heavy when stuffed into a bottle, it weighed nothing in its natural state. His attention was drawn most particularly to an old wrangle between Galileo and a Genoese philosopher named Giovanni Battista Baliani. The argument hinged on the use of siphons to transport water from one site to another, usually over a vertical barrier such as a hill. This works on the same principle as siphoning petrol from a car. Fill a long tube with water, stick one end in a pond or stream, and carry the other end over your hill. Water will then conveniently spout out of the far end, and continue to do so until you've drained the original pond or you pull the tube back out again.

Baliani had noticed that siphons seemed to have an upper height limit beyond which they didn't work. If the hill was higher than about eighteen Florentine ells (a little more than nine metres), the siphon refused to co-operate, and no water came out.

He believed that the force pushing water through the pipe was the weight of the Earth's atmosphere. Air, he said, was constantly squeezing down on the surface of the pond, and it was so heavy that it managed to push water up into the pipe. The siphon stopped operating, he reasoned, because even the weight of the entire atmosphere has its limits. At a height of more than nine metres, the air pressing down on the surface of the pond was not heavy enough to overwhelm the gravity trying to pull water back down, and the siphon would lose its power.

Galileo, however, had disagreed. Unable to believe that the atmosphere itself is heavy, he decided that the power in question wasn't pushing but sucking. On either side of the hill, he said, water was trying to fall back down out of the pipe. But as it fell, it left behind an empty space in the middle of the pipe. The complete absence of any material at all in this so-called vacuum would give it extraordinary properties, including the ability to suck. That was what drew water over the hill. If the hill was higher than nine metres, the water inside the pipe became too heavy for the vacuum's suck.

Torricelli thought that Galileo was wrong, and that the atmosphere really did push. He also decided to prove it.

First he figured out how to mimic the action of the siphon, but at a rather more manageable scale. Instead of water, he used mercury – known at the time as quicksilver, not because it moved rapidly but because it seemed almost alive. Unlike all the other cold, dead metals, liquid mercury curled itself into bright balls that darted around a tabletop and spilled on to the floor with splashes of brilliance. However, like the other metals it was also very heavy. The result from the siphons suggested that if Torricelli tried to balance the weight of the atmosphere using water, he would need a tube more than nine metres long. But with the much heavier mercury, just one metre of tube should do the trick.

So Torricelli took a metre-long glass tube that was closed at one end, filled it with mercury[6] and stopped the open end with a finger. Then he tipped the tube upside down, put it into a basin of mercury, and carefully withdrew his finger. If the air had no pressing role to play, there would now be nothing to stop the mercury from succumbing to the force of gravity and spilling back down the open tube. But if Torricelli was right, the mercury should stop at exactly the point where the weight of air pressing outside balanced its own weight. By comparing the relative weights of mercury and water, he had calculated the level at which it should stop not at eighteen ells like the water in the siphons, but at a mere ell and a quarter and a finger more.[7]

And that's exactly what happened.

But what force was keeping the mercury up? Was it the pressure of the air, or was it, as Galileo had believed, the vacuum's powerful suck? To find out, Torricelli repeated his experiment with a slight twist. He put two tubes side by side. One was a straight glass tube about one metre long and the same diameter throughout. The other was similar except that it had a large round glass globe on the closed end. Both were filled with mercury (the one with the glass globe needed rather more than the other) and then tipped upside down into the same basin.

If Galileo's argument was right, the tube with the globe on the end would have more empty space to suck with which would pull its mercury level higher. But if Torricelli was right, the mercury in both tubes should fall to exactly the same level.

The bright silver mercury slipped down the sides of both tubes and . . . ended up at exactly the same level, one ell and a quarter and a finger above the level of the bath. Torricelli was right. No matter how much vacuum was in the space above the mercury, the force holding it up was still the same. Vacuums don't suck; the air pushes.

This is a truly extraordinary notion, an effect of our atmosphere that we encounter unwittingly all the time. When you sip through a straw you may think the power of your suction pulls the drink into your mouth. But it doesn't. Your suck simply moves the air away from one side of the straw, and the drink then arrives in your mouth courtesy of the overwhelming weight of the air around you. The same thing happens when a baby drinks from its mother's breast. The baby's enthusiastic sucking just removes the air from around its mother's nipple; the force of the air above her then squeezes the mother's breast and sends milk spurting into the baby's mouth. It's the same, too, with a vacuum cleaner. The air outside pushes dust and debris up the hose because the air that had been shoving equally from the other side has now been removed. Try using a vacuum cleaner in space and you won't be picking up cosmic dust, since there isn't any air on the other side to do the pushing.

Torricelli's experiment with the glass globe had proved the weight of the atmosphere to his own satisfaction, but it would take more than that to convince the rest of the world. Part of the problem was that this notion is so counterintuitive. The air just doesn't seem to be that heavy. We can walk through it without even noticing it's there. If it really were pushing down on us continually with such a great force, why wouldn't we be crushed? (The answer is that most parts of our bodies aren't compressible, and the few collapsible spaces contain air at exactly the same pressure as the air outside. As hard as our atmosphere pushes down on us, we push back.)

It didn't help that news of the crucial experiments only trickled out gradually by whispers and rumours. Proud though he was of his findings, Torricelli didn't dare trumpet them to the rest of the world. The trouble was that he had been playing with vacuums. And the Church, in another of its unfortunate pronouncements on physics, had declared belief in the vacuum to be heretical.

The Church had decided to abhor the vacuum mainly because of the teachings of various philosophers who had lived long before Christ. Aristotle, for instance, believed that a vacuum was logically impossible. For him, space was, by definition, the place where objects resided. If there were no objects there could be no space, and hence no vacuum. The materialists Democritus and later Lucretius, however, believed that all matter was composed of tiny indivisible particles called atoms, which were separated from one another by empty space.

Not much progress was made in the following twenty-one centuries to resolve this issue, and by the sixteenth century the Catholic Church had decided to side with Aristotle. By reducing all of Creation to a collection of atoms, Democritus and Lucretius had left no room for spirit or soul, and also raised troubling questions about exactly how, scientifically speaking, the communion wafer and wine could transform into flesh and blood. Their philosophies were therefore anathema. Tarred by association, belief in the existence of a vacuum was also declared heretical. According to the religious authorities, God had decreed that a vacuum would be so unnatural that air would always rush in immediately to prevent one from being formed. To say otherwise was to risk the wrath of the Inquisition.

Having seen the effects of Galileo's mild outspokenness, Torricelli opted for discretion. He never published his results, except in one famous letter that he wrote on 11 June 1644, to his close friend Michelangelo Ricci. Though Ricci was a Jesuit, he was also a firm advocate of Torricelli's work, and Torricelli described his experiments in careful detail, with sketches of the apparatus. Mostly, he remained matter of fact, but once in a while he let his delight in his findings shine through. 'What a marvel it is!' he wrote when he contemplated the invisible air pressing his mercury up into the tube.[8] He spoke with awe of how our blanket of air, perhaps fifty miles high, constantly presses down on the planet beneath. And he encapsulated it all in this one glorious image. '*Noi viviamo sommersi nel fondo d'un pelago d'aria*,' he said. 'We live submerged at the bottom of an ocean of air.'

With his experiments in quicksilver, Torricelli had proved the pressing power of air to his own satisfaction, but his secrecy about the results and the prevailing stubborn resistance to this extraordinary new notion meant that, for the moment at least, the old ideas continued to rule.

Fortunately there remained the other person who had arrived in Florence just before Galileo's death and who, like Torricelli, was destined to pick up his mantle. His name was Robert Boyle, and when he reached Florence in October 1641, he was a sixteen-year-old schoolboy who had as yet no particular yen for science.

Boyle was the son of one of Ireland's richest noblemen. He had ridden from Geneva to Florence that summer with his brother and tutor on a leg of their Grand Tour of Europe. But unlike the other privileged young gentlemen risking pox, plague and bandits in the interest of gaining Continental polish, Boyle truly wanted to learn. He carried books everywhere; he read them walking along roads and stumbling down hillsides. He disputed philosophy and religion with fellow guests at the lodging houses and tried to make the deepest possible sense of everything he saw and heard.

Soon after Boyle arrived in Florence, he came across a copy of Galileo's final book and was deeply struck. He was also struck with indignation by the fate of the man now dying in his villa just a few miles away. Boyle noted triumphantly in his journal how, when monks went to visit the 'great star-gazer' and chided him that his blindness was a punishment sent by God, the quick-witted Galileo had replied that at least 'he had the satisfaction of not being blind till he had seen in heaven what never mortal eyes beheld before'.

For Boyle, the Church was also suffering from blindness. He decided that religion was about revealing the wonders of God's nature, not hiding them behind dreary dogma. Boyle didn't want to be told what to believe about the workings of the world. He wanted to glorify God by discovering them for himself.

Yet the seed Galileo's work had planted could easily have withered over the succeeding years. For shortly after Boyle left Florence, his home country, Ireland, erupted in rebellion, while England tumbled into its own civil war. It was more than two years before Boyle could make his way back home, and even then he got only as far as England, first to his sister's house in London and then to Stalbridge, a modest manor house that his father had bought for him in Devon.

This would have been a good time for Robert Boyle to settle into the life of a country squire. England was by then a little less troubled. True, King Charles I had been arrested, then later arraigned and publicly beheaded, but the Protector, Oliver Cromwell, had taken control and, along with his New Model Army, had restored a large measure of political stability. Boyle was comfortably off. He could indulge in gentlemanly pursuits, ride, shoot and fish.

But there was still something missing in his life. He was full of ideas but there were no obvious routes for gentlemen to express them. Boyle dabbled with religious writings. He wrote a series of 'Occasional Reflections' addressed to his favourite sister, Katherine, Lady Ranelagh, drawing what were admittedly often mawkish morals from events such as 'Upon the sighting of a fair milkmaid singing to her cow' and 'Upon my spaniel's carefulness not to lose me in a strange place'. This led to some mockery, which was not really fair. Boyle was pious, but never sanctimonious. He was pleasant, approachable and almost pathologically fair-minded, and though his religious sentiments were naive, he was still in his early twenties.

One of the most famous parodies of Boyle's 'Reflections' was penned by satirist Jonathan Swift, several decades later. Swift at the time was private chaplain to a lady who was smitten by Boyle's writings and wanted them read to her constantly. Swift became so exasperated that he slipped in an extra, unauthorised and very funny piece titled 'A pious meditation on a Broom Staff': 'But a broomstick, perhaps you will say, is an emblem of a tree standing on its head; and pray what is man, but a topsy-turvy creature . . .' (In spite of his mockery, Swift may well have made use of Boyle's vivid imagination as the inspiration for his most famous book[9]: *Gulliver's Travels*.)

Boyle even wrote a romantic yet highly moralistic novel, and for a while it seemed he might try expending his intellectual energies on a literary career. But his curiosity about the workings of the world tugged at him. He wanted to understand the world in a new way, the way that Galileo had shown him. He wanted, above all, to *experiment*.

So in 1649, Boyle installed a laboratory at Stalbridge. He

commissioned furnaces from the continent, and he dabbled with alchemical efforts to find a way to turn lead into gold. But his attempts to experiment seemed aimless. He needed to be among people who shared his urge to understand the natural world through experiment and not through reason alone. During his visits to his sister Katherine's house in London, he had met many such men, who were already discussing the best new ways to probe nature. They met in each other's homes and called themselves the 'Invisible College', though Boyle always referred to them as the 'Invisibles'. (This was the first glimmering of what would become London's famous 'Royal Society' when the monarchy was eventually restored after the death of Cromwell.) From these men and their discussions with his thoughtful, intelligent sister, Boyle had learned much. But London had begun to seem politically too unstable for these men, and many of them had moved to take up positions behind the safe walls of Oxford's rather less invisible university. And so, in the mid-1650s, Boyle decided that he would join them. He left his stately manor house and moved into lodgings that his sister found for him in the house of an apothecary.

Boyle was at last in his element. He had never been interested in the social status to which he was entitled by birth. (Nor was he particularly interested in fame or money. Throughout his life he was to turn down many offers of honours and appointments. He said with typical cleverness that he preferred to work on things that were 'luciferous rather than lucriferous', that is he preferred work that was enlightening rather than money-making.)[10] Instead, at last, he was surrounded by people who shared his passion. There were chemists and mathematicians, physicists and physicians. Here were Richard Lower and Tom Willis, who together would soon perform the world's first blood-transfusion experiment; there was Sir Christopher Wren, architect, polymath, Renaissance man. Oxford seemed full of people who were itching to experiment, to discover for themselves how the world worked.

For the first few years, Boyle watched, listened and learned. He had yet to decide what area he wanted to make his own. Meanwhile, whispers of Torricelli's experiment with quicksilver were making their way across the Continent. In France, largely beyond the reach of the Roman Inquisitors, a philosopher named Blaise Pascal had caused a great sensation with his public demonstrations, using thirty-foot-long glass tubes filled with water and wine, as well as Torricelli's preferred, but less dramatic, quicksilver. He also used the heights of the different liquids forced upward by the air to come up with a value for the total weight of the atmosphere. He announced that our ocean of air weighs some 8,283,889,440,000,000,000 pounds, and he wasn't far off.[11]

From France, news of the experiment had passed across the English Channel to London, where the 'Invisibles' were greatly taken with it and performed it many times. Even before Boyle went to Oxford, he had come across the experiment during his frequent visits to London, and it had immediately quickened his interest. He later wrote that air was the perfect subject to study. Not only is it vital for breathing, but it also touches us inside and out every day of our lives. Something that is jointly so necessary and so pervasive would surely be full of hitherto unsuspected scientific treasures. However, Torricelli's experiment had been thoroughly dissected and very frequently reproduced. There didn't seem much more that Boyle could do with it.

Then, in 1657, came sensational news. The Burgomaster of Magdeburg in Germany, one Otto von Guericke, had invented a way to pump air. His method was a little crude, but he was a terrific showman and had used his new air pump to great effect. He had taken two copper hemispheres about twenty inches in diameter, carefully milled so their edges fitted together perfectly and they formed a sealed globe, then used his air pump to remove much of the air inside the globe. Finally, he attached teams of horses, one to either side, and made them heave. With the overwhelming weight of the atmosphere squeezing the two sides together it took thirty-two straining draft horses to wrench the hemispheres apart.

Boyle was enchanted by this experiment. 'Thereby,' he wrote, 'the great force of the external air . . . was rendered more obvious and conspicuous than in any experiment I had formerly known.' It didn't quite resolve the issue. Those who were already convinced interpreted it the same way as Boyle, but it was still possible to argue that the vacuum inside the Magdeburg sphere was somehow pulling, rather than the air outside pushing.

What is more important for our story, however, is that Von Guericke had invented a new way of working with air. Before then, the only way to make a vacuum was awkwardly, at the top of a Torricellian tube full of quicksilver. Now there was a new way, one that was surely open to experiment. This was exactly what Boyle had been looking for.

Von Guericke's air pump had not been designed for the sort of experiments Boyle had in mind. There was no chamber in which to put equipment, and whatever was being pumped had to be held underwater. However, it was a start and could surely be improved upon. Boyle immediately hired Robert Hooke, the most brilliant experimental designer in England, and set him to work.

Robert Hooke was an irascible hunchback, a hypochondriac with a caustic wit and a terrifying manner. He was also a genius. As engineer and architect he would be second only to Sir Christopher Wren in rebuilding London after the fire that would destroy most of the city in just a few years. Now, although he had only recently completed his studies at Oxford, he was already renowned for his ingenuity. Hooke began to design an air pump that would do everything Boyle desired. He would have no need to fiddle around with quicksilver and thin glass tubes as Torricelli had, nor to hold his pump underwater as had Otto von Guericke. With the machine that Hooke designed, Boyle would soon be able to make air come and go at will.

While Hooke laboured, the outside world grew increasingly fearful. The stability that Oliver Cromwell had brought to

England was beginning to fray. Even nature seemed to be against him. The winter of 1657/8 was the severest on record and bitter temperatures lasted until June. There were days of public fasting to try to ward off the evil that had befallen the country. On 21 August Cromwell fell ill and the nation held its breath. Ten days later England was blasted by a storm so violent that Cromwell's followers declared it was a warning of divine retribution against his detractors, and his enemies said the devil was riding in on the wind to claim the soul of the great traitor and king-slayer. Whatever the true reason for the storm, Cromwell had only a few days more to live, and his death heralded a new period of disarray.

The Royalists began to agitate for the return of the King, while the Roundheads marshalled their forces under the banner of Cromwell's regrettably feeble son. Yet through all this, Boyle and Hooke remained oblivious. Safely ensconced in Oxford, they worked steadily on their air pump.

It wasn't easy.[12] Boyle was struggling desperately with distemper in his eyes. A few years earlier he had fallen off his horse in Ireland and contracted a debilitating illness. Soon afterwards his sight had begun to trouble him, and there were times when he could scarcely make out the apparatus for himself. But still he was eager for what he called 'the principal fruit I promised myself from our Engine'. For Boyle already believed that Torricelli and Von Guericke were right, that the driving force in Torricelli's quicksilver experiment was the weight of the air. And he also believed that, with his new air pump, he would be able to convince the rest of the world.

Boyle's idea was to take Torricelli's experiment and put the whole thing inside a vacuum. This had already been tried a couple of times before, but without an air pump it had been very messy, involving attempts to fit one glass tube filled with quicksilver inside the vacuum created at the top of another. Hooke's adaptation of Otto von Guericke's invention was going to make the experiment much easier.

At last, the pump was ready. Hooke's design consisted of a large glass globe, with a wide opening at the neck, which could hold about fifty pints. This would be the 'receiver' of the pump, where the experiments were to take place. Attached below this was a hollow brass cylinder just over a foot long, into which a plunger covered with tanned leather was tightly rammed. Through a clever system of valves, both globe and cylinder could be opened to the outside air or sealed from it. Simply pulling the plunger downwards drew air out of the globe. Adjust the appropriate valves; repeat this process several times, and you have yourself a vacuum.

The first step was to recreate Torricelli's experiment. Boyle and Hooke took a slender cylinder of glass about three feet long, closed at one end, and filled it with quicksilver. Then, as usual, they tipped it upside down into a box half-filled with mercury. Just as expected, the mercury inside the tube began to fall until it reached a height of 29½ inches.[13]

The next part was more delicate. Box, tube and all were attached to strings and let down gently to dangle in the middle of the glass globe. (The top of the glass tube still poked up through the neck of the receiver but Boyle slipped a tight cover over it to prevent any leaks.) As far as the mercury in the glass tube was concerned, nothing had changed. It still rested 29½ inches above the mercury in the box below.

Now to begin sucking air out of the glass receiver. If Galileo was right, this should change nothing. According to him, the only force holding the quicksilver up was the sucking power of the vacuum in the closed space at the top of the glass tube; the presence or absence of air in the globe outside should be irrelevant. But if Torricelli and Boyle were right, removing air from the globe would take away the force holding the mercury up and it would fall.

The assistant manning the pump grasped the handle and began firmly ratcheting it downwards. One cylinder's worth of air was drawn out of the great glass globe. And . . . the quicksilver unmistakably fell. Turn the valve, replace the plunger, try again. Another cylinder's worth of air disappeared from the globe, and the quicksilver dropped still farther down the glass tube that protruded from the top of the pump. Soon, it had disappeared below the neck of the globe and Boyle could no longer mark its level on the paper he had attached for the purpose. Hampered by his poor eyesight, he had to peer through the walls of the glass globe to make out the shiny surface of the quicksilver as, with each crank on the pump's handle, it successively lurched its way down the inside of its tube towards the box waiting below.[14]

This surely was the proof Boyle had been looking for, but to be extra careful he decided to try reversing the procedure. He turned the valve and allowed air to begin flooding back into the globe. Immediately the quicksilver raced back up the tube. The more air Boyle allowed inside the globe to squeeze down on the quicksilver, the higher up its tube it climbed. The more he removed air from the globe to take the pressure off, the farther the quicksilver fell. The pressure of air had to be keeping the quicksilver aloft. What could be clearer?

And yet the argument wasn't quite over. Boyle was now being baited by one of his *bêtes noires*, a Jesuit named Linus who was doggedly convinced that a vacuum could not exist. Instead, he declared, the answer lay in a bizarre invention of his which he called a 'funiculus'. This was some kind of strange, invisible thread that hung in the apparently empty space above the mercury, holding it up like a puppet on a string.

The mild-mannered Boyle was polite as ever in his response to this absurd idea. But even he couldn't resist saying that it was 'partly precarious, partly unintelligible, and partly insufficient, and besides . . .' – and this was the final blow – 'needless'.

For the final irrevocable proof that air really does push, and in all directions, too, was already there in the results of another of Boyle's experiments with his air pump – number 31. To do this experiment, Boyle had dispensed with the glass globe altogether. All he needed was the air pump itself.

The idea was dazzlingly simple. First, open the valve at the top of the cylinder and push the plunger all the way to the top so it fills every scrap of the cylinder's bore. Then close the valve at the top, so that no more air can rush in. Finally, attach weights to the bottom of the plunger to try to pull it back down again. 10 pounds, 20 pounds, 50 . . . 60 . . . 70 pounds. Still the plunger wouldn't budge. Finally, with 100 pounds dragging it downwards, the plunger began to fall.

By that time Boyle had made his point. There was nothing at all inside the cylinder above the plunger, no vacuum or 'funiculus' to hold it up from the inside. The force that kept the plunger in place when such a huge weight was pulling it downward must have come from the outside. It could only be the apparently insubstantial and inconsequential stuff that surrounds us and squeezes down on us every day of our lives: our all-embracing ocean of air.

Boyle published his results in 1660. By then, the Oxford group of intellectuals had largely scattered. Many of them had backed Cromwell and were now fearful of the consequences of a Royalist revival. Boyle himself had remained steadily neutral, but even he left Oxford for a while to wait out the new political uncertainties at the country house of a friend. While there, he prepared his book, to be called *New Experiments Physico-Mechanical Touching the Spring of Air.*

Though Boyle was fluent in Latin, as in many other languages, he was unusual for philosophers of the time in that he chose to write in accessible, everyday English. Still more unusually, he eschewed the 'normal' way of writing up science – philosophical discourses among fictitious persons – in favour of a straightforward description of his apparatus, what he did for each experiment, and the results he obtained. He wanted people to understand exactly what he had done, and even to be able to repeat it. In this sense, he was one of the world's first true scientists.[15]

The book was an immediate hit, not least because it contained much more than the proof of the pressing power of air. Boyle would never have been satisfied with merely confirming what Torricelli had already discovered. Armed with his new air pump, he had always wanted to go much farther.

One of the first new things Boyle discovered was that, unlike water, air seemed to have bounce. He'd noticed this almost as soon as he tried removing air from inside the glass globe. If you first pulled down the plunger to make a vacuum inside the brass cylinder, and only then opened the valve to the glass globe, air immediately whooshed from the globe into the cylinder. Everyone in the room could hear it. If you closed off the valve, emptied the cylinder and repeated the process, the whooshing still happened but it was a bit less dramatic, as if less air was rushing out of the globe. And the next time you tried, even less air whooshed out.

Boyle deduced that air must contain some kind of particles that squeeze against each other. When the globe was full of air it was like an overcrowded room; as soon as the valve was opened, particles immediately spilled out. But with each drag of the pump, the remaining particles could spread out – and were hence much more reluctant to leave.

Boyle didn't understand this quite as we do today – he imagined air to be something like a springy pile of flocks of wool. We now know that a piece of air the size of a sugar lump contains around 25 billion billion molecules all constantly darting about faster than the speed of sound. Every molecule crashes into another 5 billion times a second and it is this incessant pinball barging that gives air its spring. It's why the billions of bouncing molecules inside a tyre can hold up a truck, and why the weight of air doesn't just press downwards but acts in every direction.[16]

Boyle wanted to find out what role, if any, this springy air plays in the perception of sound. Nobody really knew how sound moves around, although there was a vague notion that it had something to do with the atmosphere.

He decided to try a careful experiment. Into his great glass globe he gently lowered a ticking watch, suspended from a thread. The watch was one of the very latest models, which had a hand to mark out the seconds as well as the more usual minutes and hours. This way, the experimenters would be able to assure themselves that the watch was still working as it dangled inside the globe.

At first, the sound of ticking was clear even a foot away from the globe. But when the pump began removing air, something changed: the ticking grew fainter and fainter. At last, when the pump had removed as much air as possible, Boyle and his helpers pressed their ears to the side of the globe. They could see the newfangled second hand as it continued to work its way around the watch face. But though everyone in the room strained to pick up the slightest hint of ticking, nobody could hear a thing. The air that left the globe had taken with it the power to transmit sound.

We now know that sound is made from vibrations. It can be transmitted through anything that wobbles – if your ear is touching something that's vibrating, you have no need of the air in between. But most of the sounds we care about happen at a distance, and for that our atmosphere is essential. Anything on Earth that makes a noise sets the air around it quivering, and our entire thick atmosphere acts like a giant vibrating drum. It's connected not by a skin but by those constant collisions in the pinball world of air molecules. You can send sounds across an entire room with just a little puff of effort, because your wobbling larynx passes on its vibrations to barging air molecules which then crash them on to their neighbours. Without air, a cannon could go off right next to your ear and you wouldn't hear, or feel, a thing. (Even the power of explosions comes from the air. When a bomb is detonated, it sends countless molecules of air flying in your direction to knock you off your feet.) Without air, our planet would be as silent as the grave.

Next, Boyle wondered just what role air played in flight. Humans were obviously earthbound and yet birds and insects had no trouble gliding through the air. Were they somehow floating like fish in the ocean above our heads? (And if so, why couldn't we float, too?)

To try to find out how air was necessary for flight, Boyle started with a humming bee. (He was a bit disappointed not to be able to try a butterfly, which seems to rely for flight more completely on wafts of air, but unfortunately the season was still too cold.) He put the bee inside the chamber with a bundle of flowers that hung from a thread near the top of the globe. He then prodded and teased the poor creature until she landed on the flowers and remained there. Next, he gradually began to draw out the air. At first the bee took no notice, and then suddenly the experiment was over. The bee tumbled helplessly down the wall of the globe without making any effort to use her wings. By the time he had managed to let air back into the chamber, she was already dead.

This wasn't exactly conclusive. Had the bee failed to fly because it had no air or because it was suffocating? Boyle tried again, this time with a lark whose wing had been broken by a hunter's shot, but which was otherwise, Boyle reported, 'very lively'. But once she was inside the chamber and losing air, it wasn't long before she, too, began to droop. Soon, she began to writhe in convulsions, throwing herself over in frantic somersaults. Hastily, Boyle's assistant turned the stopcock and let in fresh air, but once again it was too late. 'The whole tragedy,' Boyle reported, 'had been concluded within ten minutes of an hour.'

Boyle realised that his air pump was going to tell him nothing about flight. His subjects were dying before they even had a chance to flap their wings. So he diverted his attention to trying to understand breathing. What made air so vital? He wondered if an animal more used to enclosed spaces might fare better, but a mouse 'taken in such a trap as had rather affrighted than hurt him' went the way of the bird.

Observers were always welcome at Boyle's experiments, but he was now finding that rather awkward. One of his tests with another bird had to be abandoned when the subject was rescued by 'some fair lady' who was horrified by the creature's convulsions and insisted that Boyle immediately let air back in. After this, he did his more controversial experiments at night.

He began to wonder if his animals were dying because their exhalations somehow clogged up the globe. So he tried leaving a mouse in the closed vessel overnight, with a bed of paper to rest on and some cheese in case it was hungry, and carefully placed the vessel by the fire to make sure it didn't suffer from cold. Next morning, the mouse was not only alive but had eaten almost all of the cheese.

It was all very baffling. There were plenty of theories at the time for why breathing is necessary, but none of them was really enticing, which is perhaps not surprising, since none of them was right. Boyle himself inclined towards the idea that we breathe to cool down our lungs, which might otherwise become overheated. After all, cold-blooded animals such as fish have no lungs. On the other hand, Boyle correctly suspected that fish might be somehow making use of air dissolved in the water around them.

It wasn't only animals that needed air. Boyle also discovered that flames flickered out as soon as he drew air from his globe. In some cases, for glowing coals for instance, readmitting air would rekindle the flame. But if he left the coals for more than four or five minutes, the spark irrevocably died. Boyle couldn't help but notice the similarity between flame and life. 'The flame of a lamp,' he remarked, 'will last almost as little after the exsuction of the air as the life of an animal.' Air was clearly vital for both processes, but Boyle had no idea why.

At least, he remarked, the 'spring' that he had discovered made air extremely difficult to remove. Each time his pump operated, the remaining air was more reluctant to quit the globe, which Boyle decided was ultimately a good thing. 'This invited us thankfully to reflect upon the wise goodness of the creator who by giving air a spring hath made it very difficult, as men finde it, to exclude a thing so necessary to animals.'

But still, he strove to understand why. He nearly, so very nearly, came upon the answer, and his writings are full of speculations that come tantalisingly close to the mark. 'The difficulty we find of keeping flame and fire alive, though but for a little time, without air, makes me sometimes prone to suspect, that there may be dispersed throughout the rest of the atmosphere some odd substance, either of a solar, or astral, or some other exotic, nature,' he wrote once. And another time: 'I have often suspected that there may be in the air some yet more latent qualities or powers . . . due to the ingredients whereof it consists.'

That last comment is extraordinarily prescient. Nobody yet knew that air was a mixture of different gases. Even the notion of individual gases hadn't yet been invented. Air was an *element*, a pervasive substance that had no parts of its own. This was the mountain of prejudice that had yet to be overcome before the most extraordinary secrets of air would begin to emerge.

The trouble with Boyle's air pump was it removed everything at once. If the power of air came from its individual ingredients, Boyle was never going to be able to separate them out. His combination of rational mind and vivid imagination had taken him this far, but the next step continually eluded him. In the end, he turned to other things. As his eyesight grew still worse, he studied what little was known of the functions and maladies of the eyes. Ever hopeful, he seized on increasingly bizarre remedies, such as blowing powdered dung into his eyes or bathing them in honey. Once he had been able to read for ten hours a day, but now he could barely make out the words on a page.

Boyle's health continued to deteriorate along with his sight, and in 1691, aged 65, he died. In his will, Boyle bequeathed his scientific collections to the Royal Society, 'praying that they, and all other searchers into physical truths, may cordially refer their attainments to the glory of the great Author of Nature, and to the comfort of mankind'.

Like Galileo and Torricelli before him, Boyle had never married, though he always wore a mysterious ring bearing two small diamonds and an emerald. He left this ring to his beloved sister, Katherine, saying that she would know why. She, however, had died just a week earlier, and the secret died with her.

The secrets of air, though, did not die. Those three great scientists of the seventeenth century, Galileo, Torricelli and Boyle, two of them blind and one afraid of the Inquisition, had permanently changed the way we see our world. They had discovered that we live at the bottom of an ocean of air. Now, those who came after were about to discover the ways that same ocean transforms a lump of rock and stone into a living, breathing planet.

First, the answer that had perpetually eluded the frustrated Boyle. The spirit of air somehow gives life to both animals and flame. But how?

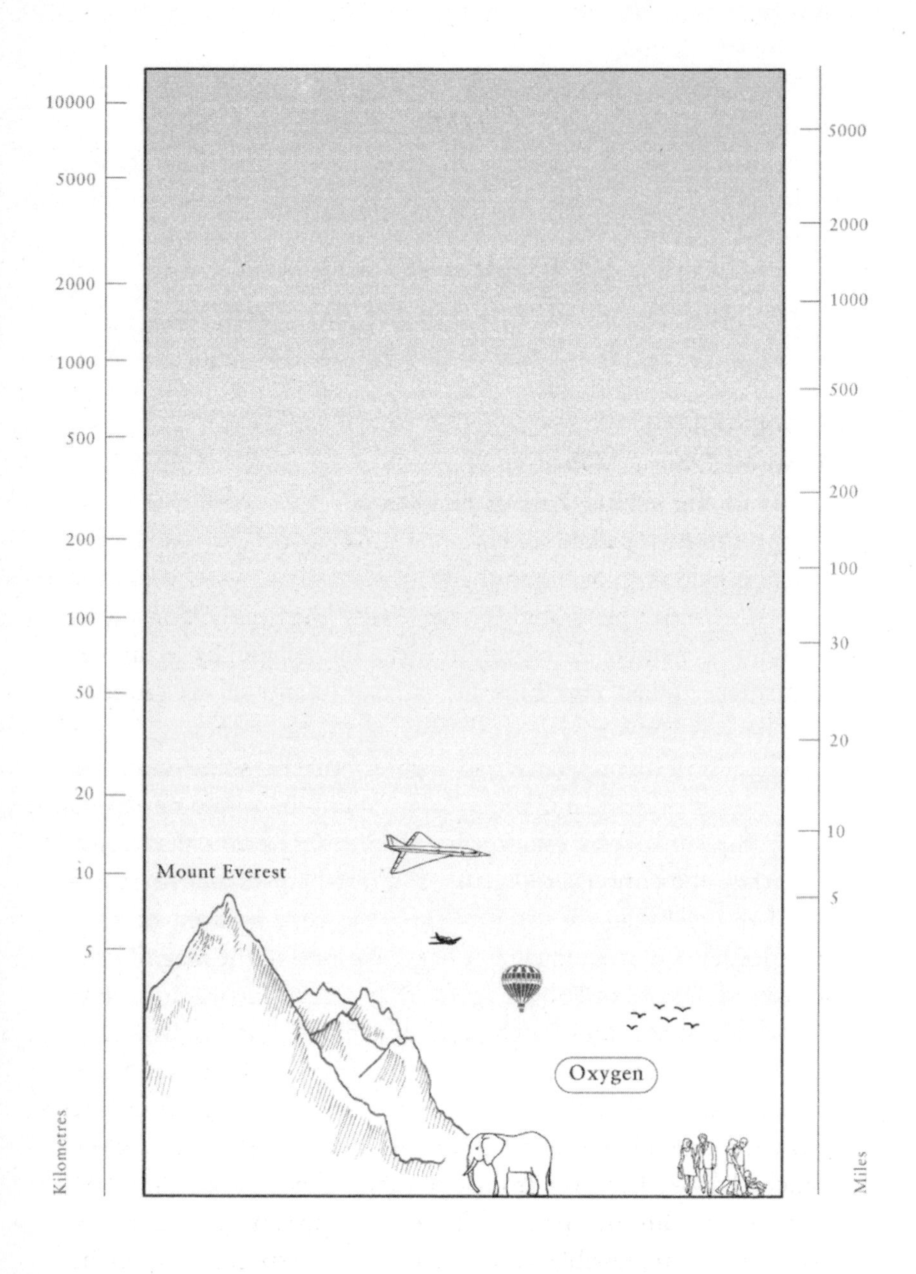

10000
5000
2000
1000
500
200
100
50
20
10
5
Kilometres
5000
2000
1000
500
200
100
30
20
10
5
Miles
Mount Everest
Oxygen

## *Chapter 2*

# ELIXIR OF LIFE

**1 August 1774**
**Bowood House, Wiltshire,**
**home of the second Earl of Shelburne**

Joseph Priestley picked up his new burning glass carefully by the rim and held it up to the sunlight. It was twelve inches wide, the size of a dinner plate, and looked like a huge magnifying glass without a handle. Ground into its lens shape by a master craftsman, it had cost him the shocking sum of six guineas. But he was convinced it would be worth the money.

The rest of his apparatus had been in place for some time, and this was the only missing piece. Now, at last, he would be able to focus the sun's light into an intense burning beam that could penetrate the curious arrangement of glass, valves and mercury-filled troughs that was assembled on the table in front of him.

Priestley was an unlikely figure to be found in the magnificent surroundings of Lord Shelburne's country seat. He was forty-one years old, of medium height and slim build. His hair was thin and unremarkable, and he rarely bothered with the carefully curled and powdered wigs that were *de rigueur* for the time. He wore the drab clothes of a clergyman. His features had a slightly pinched look, left over from childhood illnesses, and his eyes were grey. But in spite of his sober appearance, there was something irrepressible about him, and – especially when he was performing one of his experiments – an air of intense excitement. On this afternoon in particular, that excitement was more justified than ever. He was on the point of discovering something that would make him more famous than anything he had ever written and anything else he would ever do.

Joseph Priestley was born asking questions. He had been raised in a religious but Nonconformist household, and to challenge the accepted order of things was as natural to him as breathing. By the time he was old enough to take holy orders he was already asking too many questions for the religiously hidebound hierarchy of church, state and aristocracy that controlled England in the eighteenth century. In fact, he had questioned his way into abandoning so many of the basic tenets of the Anglican faith that he was banned from attending university (which was open only to those who could demonstrate a conventional belief in the articles of Anglicanism) and was summarily sacked by his first congregation.

Priestley was no firebrand, though he described himself as a 'furious freethinker'. His demeanour was pleasant, and his style at the pulpit conversational rather than inflammatory. Above all, Priestley believed in the power of reason. Throughout his life he remained cheerfully convinced that rational argument would prevail.

It rarely did. The problem was that clergymen were expected to embody the accepted order of things, not try to change it. And in Priestley's case, his attitude frequently got him dismissed. The combination of his scandalous (though gently spoken) views and his exasperating habit of trying to change people's minds meant that he rarely spent more than a few years in one place. He sometimes worked as a clergyman, sometimes as a teacher, and other times as a polemicist and prolific writer of pamphlets. By the end of his life, he'd written 150 books and pamphlets and more than one hundred papers, leading some of his contemporaries to grumble that he wrote his words slightly faster than his audience could read them.

Priestley wrote so much partly to counteract a terrible

memory. Once, for a pamphlet he was writing, he needed some details about the traditions of the Jewish Passover. Having consulted several writers and condensed the information into a shorthand paragraph, he mislaid the paper in a moment of abstraction. A fortnight later, with no memory of his previous studies, he repeated the whole exercise even down to the shorthand note. With the second note in hand, he then unexpectedly found the first, which he viewed, he says, with 'a degree of terror', believing that his mental powers had begun to fail him. But a further effort of memory made him realise that the same thing had happened before, and after that he made it a habit to write down and carefully preserve anything he didn't want to forget.

To an intellectual such as Priestley, the inability to remember would have been a distressing handicap, but it may also have been the source of part of his genius – by helping him to see the world with fresh eyes. He lived constantly in the moment. Unlike other, cooler minds, which needed quiet retreats for concentration, Priestley could work anywhere. In fact his favourite place for writing was beside the fire, surrounded by his cheerful, noisy family, with whom he would stop to exchange comments or pleasantries before continuing with his work.

Precipitancy was Priestley's trouble more than procrastination, and he was driven in all he did by an overriding curiosity. Already, the structure of grammar, the history of philosophy, theories of jurisprudence and static electricity had been subjects of his intense – and sometimes unsettling – study. 'My manner has always been to give my whole attention to a subject till I have satisfied myself with respect to it,'[1] he commented. He was also an unabashed enthusiast of learning. A true heir of the Enlightenment, he pictured knowledge spreading like a wave in all directions and believed that it would soon put an end to all usurped authority in the world. He once declared that the English hierarchy 'has reason to tremble even at an air pump'. This was the sort of proclamation that enchanted his admirers. (One even wrote a poem to laud Priestley's unfettered outspokenness: 'Champion of Truth . . . eccentric, piercing, bold / Like his own lightnings, which no chains can hold / Neglecting caution, and disdaining art / He seeks no armour for a naked heart.') However, it was also exactly the sort of statement that explained his constant trouble with employers, and which would ultimately bring about his downfall.

Priestley had no fear of false starts and misconceptions, and detailed all of his mistakes for the benefit of the 'adventurers in experimental philosophy' who would follow him. Nor was he afraid of being caught out in his errors. 'He who does not foolishly affect to be above the failings of humanity,' he once wrote, 'will not be mortified when it is proved that he is but a man.'

The latest subject to capture his attention was the hottest new topic in the world of natural philosophers. Nearly one hundred years after the death of Robert Boyle, the science of gases, or 'airs' as they were then known, had begun to take off. As well as ordinary 'common air' – the stuff that surrounds us and that we breathe – it seemed that there were several other 'airs'. A 'fixed air' that extinguished candles had been discovered to burst out of certain plants and minerals under the right conditions, and there had been recent suggestions of others, including one that exploded when exposed to a naked flame.

This was exciting news, because for centuries all the focus of natural philosophers had been on the more accessible states of matter – liquids and solids. Gases were so ephemeral and hard to study that, until Priestley's time, nobody had noticed there was more than one kind. Still, these were very early days. There were plenty of people now playing with the new gases that bubbled through their glassware, and yet none had yet realised that air itself was made up of more than one component. Any hints that common air might contain traces of separate gases were largely ascribed to impurities. In its purest form, common air was still believed to be a single element, entire and indivisible.

Priestley's interest in the new airs had begun a few years earlier, while he was living next to a brewery. He had noticed that the 'fixed air' (what we now call carbon dioxide) that bubbled out of the vats and hung above them in a choking cloud could be induced to impregnate water, to make a very refreshing beverage. In other words, he had invented soda water. 'It was with peculiar satisfaction that I first drank of this water,' he said later, 'which I believe was the first of its kind that had ever been tasted by man.' Initially he simply left a vessel of water overnight among the bubbles. Later, he developed a more sophisticated technique involving bellows. He delighted in making the fresh, zingy new drink for his friends and houseguests. He had no idea – and would probably have cared less – that his invention would eventually provide the necessary pizzazz for a billion-dollar global drinks enterprise.[2]

The problem with more sophisticated experiments with airs was that the equipment they required was expensive, out of the price range of a threadbare vicar and scholar, even one as brilliant as Priestley. However, he had recently obtained a wealthy and sympathetic patron. Priestley was often indignant when he was dismissed or his endeavours were unjustly blocked, but it rarely bothered him for more than a day or so. He sunnily assumed that something else would come along, and it usually did. The latest 'something else' had been in the form of William Fitzmaurice Petty, second Earl of Shelburne, a young, handsome and, most important, extremely rich man with a definite soft spot for revolutionaries. Both men sympathised with the struggles of the American colonists to achieve some measure of independence from their British overlords. Priestley was a good friend of Benjamin Franklin, and his writings would inspire Jeremy Bentham's famous phrase 'the greatest happiness for the greatest number', not to mention a certain other phrase that made its way into the Declaration of Independence two years later, involving life, liberty and the pursuit of happiness.

Shelburne had decided that Priestley would be an amusing addition to his household, and had invited him to take up the post of librarian for 250 pounds a year. Priestley wasn't particularly impressed by Shelburne and his wealthy, spoilt friends. 'I can truly say that I was not at all fascinated with that mode of life,' he wrote later. 'There is not only most virtue and most happiness, but even most true politeness in the middle classes of life . . . On the other hand, the passions of persons in higher life having been less controlled, are more apt to be inflamed; the idea of their rank and superiority to others seldom quits them.' Shelburne himself had a very uncertain temper, which, when taken together with the air of privilege that hung around him, meant that even his peers found him difficult. However, Priestley was never overawed, and instead considered that the lack of practice at considering others made members of Shelburne's class objects of pity rather than envy. 'On this account,' he said, 'they are readily entitled to compassion, it being the almost unavoidable consequence of their education and mode of life.'

And in truth, Priestley took full advantage of his time with Shelburne. Though the salary wasn't princely, especially as Priestley by then had a family of four to support, it was adequate. And he didn't have to bother much with stocking the library or performing mundane household tasks. As long as he was on hand now and then to impress Shelburne's guests with his latest ideas, he was free to do whatever experiments he liked. Shelburne even threw in an extra forty pounds a year for equipment, which is how Priestley finally managed to afford his much-coveted new burning glass.

This was necessary as part of an elaborate system that Priestley had devised to make and study the new airs. He knew that many solid materials would give off different kinds of air when they were heated. The problem was how to trap the airs without losing them into the surrounding 'common air'. To get around this, Priestley had invented a clever system involving a series of glass vessels. He would place some substance, in this case a solid red lump of *mercurius calcinatus* (a calx made by heating mercury in ordinary air), at the bottom of a long glass tube. Then he would fill the tube with mercury. He could then invert the tube and place it in a trough of mercury. Just as in Torricelli's experiment, some of the mercury slipped down the tube into the trough, leaving an empty space at the top, devoid of any kind of air. The only difference in Priestley's case was that now there was a small red lump resting gently on the tightly curved surface of the mercury.

Now, all Priestley had to do was heat the lump. Then he could collect any air it generated and study it. That's why he had bought the new burning glass. He would at last be able to focus the sun's rays on to the glass vessel to heat his *mercurius calcinatus* and see what happened.

He had chosen the mercury calx at random. Priestley's scientific method, like his curiosity, was both all-encompassing and chaotic. He never quite knew what would happen. In one experiment, in which he packed his materials inside a gun barrel and heated them in a fire, the gas he generated emerged so fiercely and rapidly from its prison that the whole gun barrel exploded, shattering the glass equipment that Priestley had put in place to collect its output. Fortunately, he noticed the problem just in time and leapt out of the way. Ever curious, he then repeated the experiment – explosion and all – but with an extra vessel carefully placed to trap some of the gas as it burst out. (It turned out to be what we would now call nitrous oxide, or 'laughing gas'.)

This time, however, the experiment proved a little less dangerous. Using the burning glass, Priestley carefully focused a spot of light on to the *mercurius calcinatus* and waited. Gradually, bubbles of some kind of air made their delicate way through to his collecting vessel. To Priestley's delight, this simple process produced a prodigious amount of gas. But what was it?

One of the first things to try to do, in testing any new gas, was to see its effect on a lighted candle. And when he did so, Priestley was astonished. The candle didn't immediately go out, as it would in fixed air. It didn't burn steadily and then gradually diminish, as it would in common air. Instead, Priestley's candle flared up and burned with a fierce, intense light, brighter than any he had ever seen before. What's more, it went on burning long after it should by rights have guttered out.

Priestley didn't realise it, but he had stumbled across the part of our air that gives us life: the element oxygen.[3] We now know that when anything – candles included – burns in air, it is using up the available oxygen. Priestley's 'common air' is not a single element as he and his contemporaries believed, but is instead made up of many components, of which the principal ones are oxygen and nitrogen. Dull, unreactive nitrogen comprises almost four-fifths of the volume of air, but it is present mainly as a filler. It is oxygen that provides the active ingredient, the fuel that makes up about the remaining fifth. This is what permits a candle to burn. As soon as the oxygen is used up, there is nothing left in the air but unreactive nitrogen and the candle will go out. That's why common air could support a flame for only so long, and why Priestley's new air – pure oxygen – yielded a flame that was so much brighter and lasted so much longer.

However, at the time, Priestley and most of his contemporaries had an alternative idea. They believed that a burning candle gave off a strange substance with the tongue-twisting name 'phlogiston'. The more phlogiston there was in the vessel, the harder the candle found it to force yet more of this substance into the air around it; it was like trying to squeeze extra people into a room that's already crowded. When you burned a candle in common air, it poured out more and more phlogiston until eventually the vessel was too crammed to admit any more and the candle went out.

With this doctrine firmly lodged in his mind, Priestley was baffled by the behaviour of his candle. After falling asleep with the conundrum in his head, and waking to find it still troubling him, he decided that his new air didn't contain a single scrap of phlogiston. If a room starts out empty, you can add people in a steady stream for some time before it finally fills up. In the same way, Priestley reasoned, if the new air was devoid of phlogiston, a burning candle could continue pouring out the stuff without choking its own flame. And so Priestley, who though prolific with words was not exactly stylish in his choices, gave his invention the unwieldy name 'dephlogisticated air'.

Priestley immediately began experimenting with his new air. He tried mixing it with another of the recent discoveries of the time: inflammable air, which we now call hydrogen. Inflammable air was so called because it burned readily when mixed with common air. You could even hear a slight pop as it ignited. However, Priestley discovered that if he mixed his new air with hydrogen and inserted a flame, the resulting explosion was much more impressive. Instead of a gentle pop, it sounded more like the deafening blast from a pistol. Priestley didn't know that he had discovered a most potent mixture, the same stuff that we now use for rocket fuel, but he did realise that it made a great party trick. He carried carefully sealed vials of the raw mixture around with him in his pocket to impress friends, acquaintances and in fact almost anyone who would stop and listen. He'd take out a vial, uncork it, expose it to a flame and watch their faces. The results, he said, were most satisfying: 'It has never failed to surprise every person before whom I have made the experiment.'

He also tested the effect his new air would have on a living creature, in this case a mouse. Partly to conserve mice – because it wasn't always easy to catch them – and partly out of consideration for a fellow being, Priestley did his best to keep the mice involved in his experiments alive. If he felt there was a good chance they wouldn't survive in the air he happened to be testing, when he pushed them through the water or mercury into the vessel, he kept a tight hold of their tails to pull them out as soon as they began to look distressed. And for cases where he thought the air was likely to be good for the mice, he built them a small shelf so they could rest in comfort above the water.

From previous experiments, Priestley knew that if the flask was full of common air a mouse could survive for roughly a quarter of an hour before it needed to be rescued. But when he placed his mouse in the new air, the creature continued breathing for a full half-hour before he had to whisk it back out. Though the mouse then seemed to be dead, Priestley realised that it was merely chilled, and a few moments beside the fire revived it fully.

Encouraged by this, Priestley decided, with characteristic impetuousness, to try breathing the new air for himself. He wasn't particularly afraid of the consequences – instead he relished the idea of experiencing something that, until then, had only been breathed by mice. The experience was even better than he'd hoped: 'I fancied my breast felt peculiarly light and easy for some time afterwards,' he said. 'Who can tell but that, in time, this pure air may become a fashionable article in luxury . . .'

Once again, this was highly prescient, though perhaps even Priestley couldn't have imagined that, more than two hundred years later, trendy bars across the world from Tokyo to Los Angeles to London would be offering whiffs of pure oxygen as a treatment for everything from hangovers to headaches.

Breathing pure oxygen certainly feels good, but it is not necessarily good for the health. Priestley himself noticed the 'greater strength and vivacity of the flame of a candle, in this pure air', and therein lies a warning. 'As a candle burns out much faster in [oxygen] than in common air,' he suggested, 'so we might, as may be said, *live out too fast* [Priestley's italics] and the animal powers be too soon exhausted in this pure kind of air.'

He was right that breathing pure oxygen for too long can be dangerous. Half an hour in a Los Angeles bar won't do any harm, but if you breathed pure oxygen for too long your lungs would fill with blood, and after a few days you would die. That's because the very thing that makes oxygen useful to us is also its greatest hazard. The oxygen we need to breathe is an exceptional releaser of energy. We need oxygen's reactive powers to enable us to live such vigorous lives, but even the diluted amount that we breathe in ordinary air comes with its own perils.

Priestley had other ideas for how his new air might be useful. He suggested, for instance, that strategically placed flasks could 'qualify the noxious air of a room in which much company should be confined . . . [to make it] sweet and wholesome'. But he still persisted in thinking of it as fundamentally different from ordinary air. Even the visionary Priestley maintained that 'common' breathable air was the purest possible form. That's why he was so confounded when his new air seemed even purer than the common stuff, and why he had to invent his elaborate explanation of dephlogistication to account for its properties. The doctrine of phlogiston was holding back the whole story of oxygen and its importance to us all. For Priestley, oxygen remained a curiosity, a party trick with a few potential commercial applications for right-minded entrepreneurs. To discover the vital role it plays in the life of the planet would take someone else entirely, someone who was as cool and systematic in his experiments as Priestley was chaotic, and who was prepared to think in ways that nobody had thought before.

Antoine Lavoisier was a golden child. From infancy he was doted on, the only son of a well-to-do bourgeois French household whose father was very much on the ascent. Though he lost his mother at an early age, he was raised by a childless aunt who never doubted that he was destined for greatness.

Ten years younger than Priestley, Lavoisier was born during the reign of Louis XV, a corrupt wielder of privilege who is said to have presciently declared, 'After me, the deluge.'[4] Still, during Lavoisier's youth and early manhood there must have been little sign of the revolutionary carnage that was to come, and the tragic effect it would have on his own life.

Instead, his family seemed fully favoured by fortune. In only a few generations his forebears had worked their way up from being postal couriers to attaining a considerable position in society. Lavoisier was always most precise in attitude and behaviour, and his upbringing served only to reinforce that tendency. He was raised in a household where appearances were everything and manners dictated by a delicate web of complex social codes.

Even at the exclusive Parisian school that he attended from the age of eleven, precision was reckoned a precious concept. Lavoisier's maths and science teacher was a renowned astronomer named Abbé La Caille who once spent four years on an astronomical expedition to the Cape of Good Hope, where he observed ten thousand new stars and named fourteen constellations. On his return, La Caille calculated his expenses with a level of accuracy that made many Parisians titter. He declared that the entire voyage had cost 9,144 livres and five sous, which is like totalling every single expense from four years of college down to the last few pennies.

And yet Lavoisier learned a great deal from La Caille and his other teachers. It didn't take them long to realise that they were dealing with an uncommon talent. To be sure, Lavoisier's grasp of the humanities was shaky – he never succeeded in mastering languages, and his understanding of art was technically appreciative rather than intuitive. But in maths and science he excelled, and his teacher's encouragement fed his own natural ambition. He became determined to discover something truly remarkable. 'I am young,' he wrote, 'and avid for glory.' He dabbled in geology, astronomy and the mysteries of the weather as he cast around for something that would make his scientific name.

The behaviour of Lavoisier's adoring family did little to sap his supreme self-confidence. Once, as a young man, he had been accompanying a geologist named Guettard on a research trip for several weeks when his father offered to drive out and meet the travellers at a small town on their way home. Splendid, replied Lavoisier, and could he please bring along a bowl of goldfish as a gift for the lady with whom they had most recently been staying. At this, even Lavoisier's infatuated father was taken aback, protesting that he would have to carry the bowl in his own arms the whole way while the water slopped this way and that in a lurching carriage. However, he brought the fish.

Lavoisier was undoubtedly arrogant, but he was also fair, at least when compared with the abhorrent corruptions of the time. In 1767, at the age of twenty-four, he used a family legacy to further his financial fortunes by buying a share in the notorious Ferme Générale ('General Farm'). France at the time was run under an iniquitous system of taxation. Peasants were forced on pain of deportation to the slave galleys to pay ruinous duties even for necessities such as salt, while the wealthy paid nothing. All indirect taxation on materials such as salt and tobacco was administered by a shadowy group of people called the 'Farmers General', though they themselves had little to do with farming. Instead, as long as they delivered the requisite amount of money to the king, this body was free to charge the unlucky peasants as much indirect taxation as they wished. 'Those who consider the blood of the people as nothing in comparison with the revenue of a prince,' the great economist Adam Smith later commented, 'may perhaps approve of this method of levying taxes.'[5]

Lavoisier made a great deal of money from his involvement in the Ferme, but he abhorred its unfairness and did his best to leaven the worst injustices. Among other achievements, he managed to abolish the 'tax of the cloven hoof', whereby any Jews wishing to pass through a certain region had to pay thirty pieces of silver.[6] He also ensured that all his own dealings were as honest as the system could allow.

But although he disliked the system of taxation in part for moral reasons, he was at least as troubled by the absurd inefficiency of overtaxing some people almost to the point of extinction while letting others off completely. Any sort of inefficiency pained him. Lavoisier dealt with his financial affairs with the same careful precision that he was to accord his scientific experiments: unlike most of his compatriots in the Ferme, he recorded every transaction and accounted for every sou.

Lavoisier's work with the Ferme took up almost all his time, but very little of his creative energy. He was still fired with an ambition to achieve something more remarkable than just making money, and he began to work exhaustively to find a subject worthy of his scientific attention. He conducted his research from six to nine each morning and seven to ten each evening, and in addition devoted one full day a week, his '*jour de bonheur*' ('day of happiness'), to his favourite activity.

Meteorology held his attention for a while – he had been taking daily barometric measurements for several years and continued to do so for most of his life – but it didn't have quite the spark he sought. Then, after performing an expensive experiment to prove that diamonds are combustible, Lavoisier began to wonder why some materials burn while others don't.

He was aware of the prevailing theory of phlogiston, but he wasn't convinced by it. To most natural philosophers of the time, Priestley included, phlogiston was a very sensible concept. If you watch something burning, it's easy to believe that the flames are releasing some substance from inside the material, and that the more of this substance – phlogiston – it contains, the more easily it burns.

However, Lavoisier was troubled by the undeniable fact that when many substances – iron, for instance – are heated in air they become not lighter, but heavier. Until then, theorists had fudged the answer to this mystery by declaring that phlogiston must have some kind of negative weight, so that losing it makes you heavier. To Lavoisier, that seemed like nonsense. If something gains weight when it burns, he reasoned, it must surely be absorbing rather than releasing something. The question was, what?

To try to find out, Lavoisier began to study the work of every natural philosopher he could find who had worked on the problem, including that of Priestley. Lavoisier spoke no English, but his young wife was very proficient at languages and spent much of her time translating for her husband. She had every reason to be grateful to him. At the age of fourteen she had been troubled by a proposal of marriage from a wealthy and powerful man in his fifties who had seemed to her like an ogre. Lavoisier, then aged twenty-eight, who knew her father and was already fond of her, had rescued her from a horrible fate by the simple expedient of marrying her himself.

Lavoisier was impressed with the breadth of Priestley's investigation. He described it as 'most painstaking and interesting work'. But he was disgusted by Priestley's chaotic style of investigations, and the way that he raced from one topic to another with little thought as to what might connect the whole. Priestley's work, said Lavoisier, 'consists more or less of a web of experiments, almost uninterrupted by any reasoning'.[7]

And that was where Lavoisier saw his chance. He knew his brain was at least as brilliant as the chaotic and passionate Priestley's. But Lavoisier had something else as well: the cool head and precise habits of a financier. Put these together and he could achieve what nobody had done before. He could find out not just what happens when something burns, but why.

So, shortly after his marriage, Lavoisier began a series of meticulous experiments. First, he confirmed what he and everybody else already knew. He carefully weighed various materials such as phosphorus and lead, burned them in common air and measured the weight of the ash that was left. Every time he tried this, the ash was heavier than the material he had started with. This was just as he had expected.

However, Lavoisier's next experiment was much more ingenious. He placed some lead on a balance inside a glass jar full of air, which he sealed. Then he carefully weighed the entire jar, lead, balance and all. Next, he heated the lead from the outside and watched as the balance gradually tipped to show the lead gaining its weight. Finally – and this was the clever part – without opening the jar he weighed it again. Even though he could look through the glass walls and see from the tipped balance inside that the lead had grown significantly heavier, the weight of the entire jar remained exactly the same. Whatever had caused the lead to gain weight must have come from inside the jar.

It seemed unlikely that the extra weight had come from the glass walls or the balance. The most obvious source was the air. But how to prove it? Lavoisier reasoned that if some of the jar's air had disappeared into the lead, it must have left a gap, a partial vacuum waiting to be filled. So he unsealed the jar and, sure enough, air from the outside rushed in to fill the gap. And then he weighed the vessel again to see how much new air had entered. The answer: precisely the same amount as had disappeared into the lead.

It was in the very precision of his measurements that Lavoisier had started to find his answers. Many people had burned one material with another, weighed them in a desultory fashion and surmised what might be happening. But Lavoisier of the tidy mind and precise habits was the first to put it all together into a quantitative whole. The lead increased in weight by this amount. The air above it lost this quantity. Since the two values are precisely the same, a portion of the air must have gone into the lead. And since the remainder of the air turned out to be incapable of supporting further burning, the missing air – about a fifth – must be different from the rest.

This was momentous news. Lavoisier had discovered that common air was not a single indivisible element; instead, it was clearly a mixture of other things. And one of them, making up about one-fifth of its bulk, was the mysterious and powerful substance that allowed materials to burn and combined with them in the process.

But, frustratingly, Lavoisier still didn't know what this substance actually was. He could watch it disappear from common air but couldn't make it reappear. Once lead has burned and taken up its oxygen, it won't release it again no matter how much you heat it. Lavoisier managed to make the lead ash and other calxes yield fixed air, by burning them with charcoal, but he couldn't retrieve the exact gas that they had taken up from common air in the first place. He needed to get at the air trapped inside his lead in order to release it and study it and discover what it was, but it remained stubbornly locked away.

Lavoisier knew he needed to find another material, one that would soak up the mysterious ingredient from the air when it was heated but would release it afterwards. Lead wouldn't do this, nor would sulphur or tin or any of the other materials Lavoisier tried. For the moment at least, he was stumped.

Then, in October 1774, Lavoisier learned that Priestley himself was in town. Priestley was in the middle of a tour of European countries with his patron, Lord Shelburne, and Paris was their latest stop. Priestley wasn't much impressed with the city. Though its buildings were undoubtedly beautiful, parts of the city remained positively medieval. Foul-smelling open sewers ran down the centres of what, some hundred years later, would become the city's elegant boulevards, and there were none of the pavements that already graced London's streets. With a provincial Englishman's disregard for foreigners, Priestley also decided that many of the people he met were 'too much taken up with themselves to admit of that minute and benevolent attitude to others, which is essential to politeness'.

In spite of these criticisms of the Parisians and their habits, which might have had more to do with his indifferent French than any real impoliteness, Priestley was lionised throughout the city. Though news of his experiment with *mercurius calcinatus* had not yet filtered out, since he had performed it just a few months earlier, his previous work on the new airs was known throughout Europe and he was already famous. By now Lavoisier was considered France's foremost natural philosopher, and a meeting between the two was inevitable. Thus, one evening that autumn, the Lavoisiers invited Priestley to dine at their house along with most of the city's resident intellectuals. And naturally enough, during the course of the evening, stammering in poor French that was occasionally supplemented by Madame

Lavoisier's helpful translation, Priestley told Lavoisier about his experiment.

He told how he had made the *mercurius calcinatus* by burning mercury in air until the silver liquid turned into a crumbling red powder, and then how he had trapped this powder in his tube of mercury and heated it with his precious burning glass until it spewed out a mysterious new air that caused candles to burn with a dazzling, incandescent light. It was almost as if the *mercurius calcinatus* had trapped within it the essence of fire.

Lavoisier was riveted. Could this finally be the material he was looking for? When Priestley left, he dropped his useless lead and tin and started work on *mercurius calcinatus.*

First, Lavoisier took four ounces of very pure mercury and put it in a closed glass vessel with fifty cubic inches of common air. Then he heated it almost to its boiling point and kept it that way for twelve days. At the beginning, nothing much happened. But gradually red specks began to appear on the mercury's silver surface, and they grew larger each day. By the end of the twelve days, the reaction seemed to be at an end. Lavoisier had lost nine cubic inches of air and gained forty-five grains of red *mercurius calcinatus.* The air left behind in the vessel would not permit a candle to burn, but unlike fixed air it did not turn lime water cloudy. This was some other form of air that apparently existed only to dilute the vibrant, active part.

With the utmost care, Lavoisier collected the forty-five red grains and put them in a small glass jar whose long, thin neck twisted around itself several times and then poked up into a bell jar full of water. Now all he had to do was heat the grains of *mercurius calcinatus.* As he did so, out and up bubbled the very air they had trapped within them. Exactly nine cubic inches made their way into the bell jar above. As a final proof, Lavoisier took this air and recombined it with the stuff that had been left behind from the first experiment, the stuff that would not support burning but would neither turn lime water cloudy. Immediately this mixture became indistinguishable from common air. Candles burned normally in it; animals breathed happily for exactly as long as you would expect.

Lavoisier had found the magic ingredient, the active part of the air. He had extracted it, trapped it inside mercury, released it and recombined it with the passive part to regenerate common air. By applying his painstaking system of accounting to science, he had looked into the heart of a flame. He now knew what fed every fire on Earth.

But what to call it? Lavoisier had no patience with Priestley's name for this new gas, 'dephlogisticated air'. His experiments had clearly proved that burning had nothing to do with phlogiston and everything to do with the presence or absence of this one crucial active ingredient. Instead, since it seemed to be trapped in many different kinds of acid, he named it 'oxy-gene', which means 'acid-born'.

Lavoisier was intrigued with his new gas and began to work on it in earnest. In particular he wanted to know more about the relationship between burning and breathing, and the role that oxy-gene might play in each. Like Priestley, Lavoisier had noticed the similarity between these two processes. Place a burning candle in a closed jar of common air and eventually the flame will sputter and die. Place a living mouse in such a jar and after a while the animal will no longer be able to breathe. To Priestley, both candle and mouse were giving out phlogiston. To Lavoisier, both were using up oxy-gene. And now, he wondered how far the similarity between the two processes went. How could the same substance that fed a flame also feed life itself?

Until now, nobody had made any truly systematic investigations into the nature of breathing. Obviously, it was necessary for life. And just as obviously, food somehow sustained life. But there was no sense that food in a person was like fuel in a machine. Aristotle had believed that the purpose of breathing was to cool the blood, and this was still a popular notion even in Lavoisier's time. Other philosophers thought that breathing in a confined space became increasingly difficult because it reduced the elasticity of the air, which prevented it from pushing back enough to inflate the lungs properly. As to what relationship this had to eating, nobody really knew.

So, Lavoisier began his experiments. Unusually for him, he performed them with a collaborator, a young mathematical genius named Pierre-Simon Laplace. Among his other achievements, Laplace would later produce the complex equations that govern the behaviour of the solar system, and it is sometimes said that his efforts in this regard were halted because his equations were so successful at accounting for the available facts that, until more observations could be made, there was nothing left to explain.[8] Laplace was already famous, the most talented mathematician in the known world, and together he and Lavoisier devised a series of experiments to understand the nature of breathing.

For their experiments they used small hairy rodents lately returned from the jungles of South America. These 'guinea pigs' were very convenient in the laboratory, wrote Lavoisier, because they were 'tame, healthy creatures, easy to feed and big enough to inspire and expire air in quantities suitable for measurement'. Lavoisier had designed a clever piece of apparatus to discover the relationship between the amount of oxy-gene these guinea pigs consumed and the heat they gave off. The heat was the hard part. Lavoisier had decided to measure this by the melting of ice. He made a large, sealed, circular chamber comprising three concentric rings. The innermost ring contained the guinea pig, the second ring was packed with a known quantity of ice, and the outermost ring was filled with snow to prevent the heat of the room from reaching the ice and melting it. Lavoisier and Laplace set out to monitor what happened first when the guinea pigs were at rest, and then when they became steadily more active.

The results, enlivened by some complex equations from Laplace, were exactly what Lavoisier had hoped for. The more work the guinea pigs did, the more oxy-gene they used, and the more heat they gave off. Lavoisier was convinced. 'Respiration is a process of combustion,' he wrote, 'which, though it takes place very slowly, is perfectly analogous to the combustion of coal.' In the same way that coal supplies the fuel for a fire, some derivative of food must also provide the raw material for the energy by which we all live. And as oxy-gene feeds the glowing flame, so, too, must it release the energy stored somewhere inside us.

Lavoisier had discovered something truly important. Flames do indeed use oxygen to generate energy from candles or wood, and he was right that when we breathe we're using it to burn our food in much the same way. That's one reason we talk about 'burning calories'. If this sounds dangerous, it is. As Priestley had suspected, and Lavoisier had now begun to prove, breathing oxygen is what allows us to live such vivid, active lives. But we pay a heavy price, for oxygen is also the reason why we grow old and die.

All living things need to breathe. That is, they have to generate energy when they need it from the food reserves stored in their bodies. In our case, we have reserves of sugar, protein and fat, which sit around inside us like a pile of logs waiting to be burned. Every breath we take uses oxygen to convert some of those reserves into the energy we need to move, stay warm and do everything else that makes us human.

But oxygen isn't the only chemical that living things can use for breathing. Indeed the bacteria that constituted the first life on Earth were forced to use something much less efficient, for the simple reason that, when the world first formed more than 4.5 billion years ago, the atmosphere contained no oxygen at all. Oxygen didn't appear in the atmosphere for more than 2 billion years, and it finally showed up only because of a dramatic but inadvertent case of planetary pollution. Without that accidental air-spill, there would be no life on Earth bigger than a pinhead.

When the planet was born, it came blanketed in an ocean of air. Like the sun and the other planets in our solar system, Earth was formed when a shapeless cloud of gases, dust and fragments of rock began to collapse and coalesce. The rocks and dust trapped some of the gases between them like mortar between the bricks, and much of the rest settled on the outside of the planets in a shroud held in place by the power of gravity.[9]

This early ocean of air was just as dense as today's, and it would have looked very similar. But the lack of oxygen made a big difference to Earth's surface. The rocks, for instance, were a uniform dull grey colour – without oxygen, the iron they contained couldn't rust to the lovely reds and ochres that we see today. Still, the young Earth wasn't without its beauties. The skies periodically shed a gentle yellow rain of elemental sulphur, and the earliest beaches sparkled with golden iron pyrite. Also known as 'fool's gold', this exists today only deep underground, safely away from the oxidising air, where its vibrant colour can still confuse naive miners on the hunt for nuggets of the real thing.

For animals like us, this early atmosphere would have been an impossibly suffocating place, but the world's first occupants had an alternative way of releasing their energy. Instead of oxygen, they 'breathed' the gas that Priestley and his contemporaries called 'inflammable air' and that we call hydrogen. In the process, they made methane – 'natural gas'. Since this wasn't nearly as efficient as using oxygen, the creatures breathing it couldn't grow large. Instead, they remained the way they had begun, as microscopic pinpricks in the fabric of life.

So it was, and so it would always have been if not for a new chemical reaction invented sometime between 2.5 and 3.5 billion years ago by microbes called cyanobacteria. These creatures are so tiny that a droplet of water can contain billions of them – as many as there are people on Earth. However, they are also ubiquitous. Today you can find them in drainpipes, puddles or anywhere water is left to stand for a while and starts to go that distinctive green colour that shows they are working their magic. For they are the microbes that learned how to use the sun's energy to split water and make food, in a process we now call photosynthesis. And in doing so, they give off delicate bubbles of a certain waste product: oxygen.

This is the reason that we can breathe today. Cyanobacteria and the green plants that later incorporated their invention are now part of a giant enterprise that acts as Earth's lungs. As fast as we animals use up oxygen by breathing, plants return it to the atmosphere. It's almost as if living plants are working to make the world habitable for us – as if the most important component of our atmosphere has been made by life, for life.

(Oxygen didn't actually appear in the atmosphere until several hundred million years after it was created in this way as a by-product of photosynthesis. At first, it reacted with Earth's rocks and oceans as quickly as it was made. In the case of the oceans, the dissolved iron they contained turned to rust and fell to the sea floor, making vast mountains of debris that have turned into the world's biggest iron mines. Whenever you eat with a stainless-steel fork, or drive a car, you're probably benefiting from this early rain of rust.)

Oxygen is fantastically reactive. When it engages in chemistry, it can release large amounts of energy, which can in turn be used to fuel the activity of living things. So, oxygen's arrival in the air had a dramatic effect on the course of evolution. As long as there was too little oxygen in the atmosphere to be useful to the creatures below, they were forced to remain both sluggish and microscopic. For billions of years, the planet was coated with nothing more than primordial slime.

But gradually, inexorably, more and more oxygen trickled into the sky until one point, nearly 600 million years ago, when the atmosphere tripped over its oxygen threshold. The result was the most dramatic burst of evolutionary change in planet Earth's history. Huge new creatures suddenly appeared, some of them more than a metre long. They weren't only big; they were inventive, and almost unbelievably varied after the dull slime that came before.[10] These new creatures had shapes. They had eyes and teeth and legs and shells. They had learned to make their bodies out of not just one cell, but many. They were the world's first animals.

It's hard to overemphasise the importance of this evolutionary step. Think of the transition between cottage industries and the Industrial Revolution. Before this point, a single cell had to do everything that life needs – eat, excrete, breathe, reproduce, all in one tiny sac. Afterwards, cells could specialise and share the load. Some became arms, some hair, brains or bones. Creatures were no longer restricted to the size of pinheads. What's more, they had muscles to drive their new bodies, and that meant that at last they could move. Imagine a life without moving, and the difference it makes when suddenly you can. The new earthlings could seek out new sources of food, including other creatures. Some could chase and others could flee. They developed armour to protect themselves and weapons to attack. They learned new skills, took on new shapes and colours, and ultimately became the vivid variable life-forms we see around us on Earth today, including humans.

Nobody knows exactly the mechanism by which that final rise in oxygen triggered the appearance of animals,[11] but what's certain is that there could be no complex life without it. To be big and multicelled requires huge amounts of energy, and it takes oxygen to generate that sort of power. Every other way of breathing is simply too feeble. We need oxygen because we need its spectacular reactivity. Without it, humans could never have existed.

This reactivity comes with its own dangers. As Priestley suspected when he saw how brightly candles flared up in his new gas, breathing oxygen is like playing with fire. And we are gradually, all of us, getting burned.

That's because whenever oxygen gets involved in chemical reactions, it releases tiny negatively charged particles called electrons. All atoms and molecules contain these particles and, like people, they are most stable when they are in pairs. A chemical entity that contains one of these single, footloose electrons is called a free radical, and it is one of the most reactive, and destructive, forces on the planet. Free radicals rip through everything in their path, splitting apart stable pairs and creating yet more free radicals, which head off on destructive paths of their own. That's what happens, for instance, if you are exposed to radioactivity. The damage comes not from radiation itself, but from the free radicals it generates.

And the trouble is that when we use oxygen to breathe, there are always some electrons that break free. Even if you're doing nothing but breathe, about 2 per cent of the oxygen you consume still escapes as free radicals. If you're exercising vigorously, it's more like 10 per cent. According to one calculation, the potential damage from simply breathing for one year is equivalent to the radiation from ten thousand chest x-rays.[12]

When oxygen first appeared on Earth some 2.2 billion years ago, it was certainly a deadly poison for many of the earliest microbes. The methane-producers simply couldn't cope with the free radicals that were suddenly ripping through their bodies, tearing apart their vital chemicals. To survive, these organisms had to find refuge. They persist today in places that are comfortably moist and yet are hidden from the probing fingers of the atmosphere. That's why paddy fields give off methane, why swamps yield marsh gas that sometimes ignites into the ghostly dancing flames of legend, and why animals, including humans, generate natural gas in their guts. We fart because our intestines are now airless sanctuaries for those poisoned earthlings.

Other organisms, the ones from which we're descended, developed various complex strategies to deal with the worst depredations of oxygen. In particular, our own bodies are permanently ready to deploy an army of chemicals called antioxidants. A full-scale war is taking place every second, in every cell of our bodies, to stop the free radicals from forming, mop up the ones that do, or commit cell suicide if the invading forces become overwhelming. But the powerhouses inside our cells spend their lives playing with fire, and the long, slow leakage gradually wears us down. All the diseases of old age – dementia, cancer, heart disease – come from the accumulation of damage caused by escaped free radicals. That's one reason that eating fruit and vegetables helps protect us from these diseases – both are packed with antioxidants that help to mop up free radicals.

It's also why smoking brings these diseases on earlier in life than we would otherwise expect. Nicotine isn't the problem, except insofar as it's addictive and so encourages you to smoke more. The real damage comes from the smoke itself, which is crammed with chemicals that react with oxygen to generate bucket-loads of free radicals – something like a million billion in every puff.[13]

So can we somehow stave off old age by consuming more antioxidants? It seems not. In spite of the evident benefits from fruit and vegetables, there's no clear sign that eating 'antioxidant supplements' bought from the health-food section of your local grocery have the same beneficial effect. In fact, too many packaged antioxidants might hurt, rather than help. It takes us so long to grow old because our bodies have evolved such careful strategies to protect us from the worst effects of free radicals. Eating extra antioxidants might interfere with these natural mechanisms, like unruly mercenaries disrupting the operation of a highly trained army.

The damage from oxygen is also one reason humans are made up of two sexes. Every cell in our bodies possesses tiny powerhouses called mitochondria, which are the location where all the oxygen-burning takes place. These mitochondria are at the forefront of all the free-radical damage, and it's crucial to make sure that the ones handed down to the next generation are free from the damage that comes with ageing. A woman's eggs are born with her, and spend their lives using essentially no energy at all. Their mitochondria are kept in pristine cold storage, ready for the children to use, which is why eggs sit and wait to be fertilised rather than going off in search of a sperm.

Meanwhile, every time a man's sperm are regenerated, the mitochondria contained in the new ones are a little older. They also use plenty of energy to swim along and find the stationary egg. But after that – and here's the clever part – the mitochondria from the sperm get jettisoned, like spent rocket stages. So every child inherits pristine mitochondria from its mother, and the ageing clock doesn't start ticking until the foetus starts to form. If we had only one gender to play with, this couldn't happen. Hence the troubles, and glories, of romantic relationships between men and women are born in the chemistry of oxygen.[14]

The lesson of oxygen shows that many things that are exhilarating have their own attendant dangers: making discoveries, making enemies, challenging the authorities, falling in love. Indeed, everything about our vigorous, dynamic lifestyles comes with its own terrible cost. For our agile minds, our strong bodies, our different sexes, for the power of movement itself, we have to accept the inevitability of old age and death. The oxygen in each breath you take brings you everything that's worth living for, but it will ultimately make you pay with your life. Within its chemistry lies the very heart of the human condition.

Lavoisier didn't know the extraordinary ways in which oxygen has shaped our world and our lives, but he did know that he had proved that the most essential and vibrant ingredient for life comes from the air. He had also discovered that we breathe to burn our body's fuel, something that came as a great surprise to scientists of the eighteenth century. Until that point, eating and breathing were considered wholly unrelated activities. And this led the fair-minded Lavoisier to an uncomfortable conclusion. 'As long as we considered respiration simply as a matter of consumption of air,' he wrote, 'the position of the rich and the poor seemed the same; air is available to all and costs nothing.'

Now, however, it was clear that when people worked harder and hence breathed faster, that meant they were also burning up more of their body's food. 'By what mischance,' he demanded, 'does it happen that a poor man, who lives by manual work, who is obliged, in order to live, to put forward the greatest effort of which his body is capable, is actually forced to consume more substance than the rich man, who has less need of repair? Why, in shocking contrast, does the rich man enjoy an abundance which is not physically necessary, and which seems more appropriate to the man of toil?'[15]

This was a good time to ask. Lavoisier and Laplace published their findings in a *Memoir to the Royal Academy of Sciences* in 1789, the year of the revolution. The reign of Louis XV's feeble grandson and heir, Louis XVI, had come to a violent end. The Bastille had fallen, and Paris was replete with the promise of change. Filled with optimism, convinced that his beloved France finally faced the prospect of real reform, Lavoisier wrote that one should not chastise nature for the 'shocking contrast' that he and Laplace had uncovered. 'Let us rather rely on the progress of philosophy and humanity, which unite to promise us wise institutions, which will tend to equalise all incomes, to raise the price of labour, and to ensure its just reward; which will obtain for all social classes, and particularly for the indigent, pleasure and happiness in greater abundance.'[16]

Over in England, Priestley continued to bicker with Lavoisier by mail about the existence or otherwise of phlogiston, but he, too, was thrilled by the fall of the Bastille. Taken together with the newly independent America, Priestley decided that the world faced 'a most wonderful and important era in the history of mankind'. 'We may expect,' he went on to declare, 'to see the extinction of all national prejudice, and enmity, and the establishment of universal peace and good will among all nations.'[17]

If only their political instincts had matched their scientific ones. For as well as all those attributes that had enabled them to see in air what nobody had seen before – self-confidence, a certain tactlessness, fearlessness, refusal to take anything at face value, and a fervent curiosity – Lavoisier and Priestley had another thing in common. The French Revolution, which they each hailed with such delight, was about to destroy both their lives.

**14 July 1791**
**Birmingham, England**

Two years to the day after the fall of the Bastille, the second anniversary of the beginning of the French Revolution, Joseph Priestley and his friends were preparing to celebrate. Priestley was no longer in the employ of the Earl of Shelburne. Eventually his outspokenness had begun to embarrass the earl, who enjoyed the odd bit of revolutionary zeal but had realised that his own political ambitions were beginning to suffer from Priestley's diatribes. So Priestley had once again been forced to move, this time to Birmingham, to a house provided by his brother-in-law.

He wasn't too bothered by the change. Shelburne was still paying him a pension, and other patrons had stepped in to supply him with first-class scientific equipment to pursue his fascination with air. He had his library, his writing, his family around him, and the priceless company of other equally curious and indefatigable intellectuals: James Watt, inventor of the steam engine; Erasmus Darwin, grandfather of Charles, whose own declared aim was to 'inlist Imagination under the banner of Science';[18] and Josiah Wedgwood, innovator with ceramics and founder of the famous china company that persists in England today, who was also the grandfather of Charles Darwin, since his daughter married Erasmus's son. These friends and enthusiasts met to discuss their ideas once a month, when the moon was full, to enable them to see their way home afterwards. Thus they called themselves the Lunar Society, though when their ideas were at their wildest others were known to change that to 'Lunatics Society'. Priestley basked in the approval and stimulation of his newfound friends and sympathisers, and in some respects had never been happier.

He was also freer than ever to preach his uncomfortable, dissenting views. He had found a new position as minister in a local church, and divided his time most contentedly between studying air and writing about ways to ameliorate the lot of mankind. In particular, he continued to glory in the recent American and French revolutions, which he felt demonstrated the beginning of the triumph of reason and merit over inherited elitism.

However, these events abroad were sending tremors of alarm through the English hierarchy, and a tide of patriotism was sweeping the country. In a climate such as this, it was tactless of Priestley to describe the prevailing Establishment of church, monarchy and aristocracy in England as a 'fungus', which, he said, was sapping the juices of the country on which it parasitically fed. His was not a popular opinion among the loyal townsfolk of Birmingham, to whose ears Priestley's call for rational argument sounded more like treason. He was already used to seeing 'Damn Priestley' scrawled up on the walls, and having small boys follow him in the street to repeat what their elders had evidently taught them. It didn't bother him much, because he was cheerfully conscious that it was undeserved.

But when his friends organised a dinner in honour of the French Revolution, rumours began to circulate that Priestley had called for the head of the king, and threatened to bomb the established Anglican churches, and later that same evening a rabble was roused. In an uncharacteristic burst of prudence, Priestley hadn't even attended the dinner. He was at home playing backgammon when there was a violent banging on the door and young men, breathless from running, stammered out their news. A mob had already smashed the windows of the hotel where the dinner had taken place and torched the church where Priestley preached. It was now heading for his house, hell-bent on murder.

Priestley couldn't credit that he himself might be in danger. Who, after all, would want to hurt someone so evidently harmless? But he accepted that he might be subjected to some unpleasantness if he stayed, and he agreed to remove temporarily to the house of a neighbour. He went calmly upstairs, put a few papers and things of value where he thought no miscreants would find them, and left in the clothes he had been wearing. He told the servants to lock the doors, and if stones should be thrown to keep out of the way of the windows.

Priestley's son was less phlegmatic. He urgently did his best to secure the house, and doused every fire and every candle. At midnight the mob arrived. The weather was calm and clear, and from his neighbour's house barely a mile away Priestley could hear every shout and curse, every stroke of the implements used to break down the doors and windows. Next came the splintering sound of smashing furniture, followed by the shattering of glass. With increasing horror, Priestley realised that they were not just breaking the windows; they had started on his scientific instruments. His beloved laboratory had been one of the best-equipped in Europe. Now, helplessly, he was witnessing its destruction.

Worse was to come. They were seeking a fire. They wanted to torch his library! Priestley listened in anguish as the mob sought, at first in vain, for a flame. Somebody shouted, offering a full two guineas for a lighted candle. Priestley thought of the diaries that he had been keeping almost every day for the past forty years, each with its own record of his state of mind, his hopes and intentions and prospects for the year ahead. He thought of the many notebooks containing the fruits of his reading almost since he had first learned to formulate his opinions. In his library, too, was every sermon he had ever written; his memoirs, which were to have been published after his death; every letter he had ever received, from dear friends and learned foreigners.

And he thought, too, of his books. It had been his custom to read always with a pencil in his hand, marking the passages that he wished to look back to, or that he thought would be particularly useful for some new endeavour. He would make an index to such passages on a blank leaf at the end of the book. His library, his precious library, contained not just books, but the chief fruit of his labour and judgement in reading them. The fate of all this rested now on whether or not the mob despoiling his house could lay their hands on a flame.

And then, from somewhere, somehow, they found one. What was first an orange glow reddened as the fire took hold. It burned every bit as brightly as the candle that had enchanted Priestley when he first thrust it into his dephlogisticated air. Fed by copious amounts of the oxygen that had already made him famous, its flames ripped through the pages on which he had explained their secrets. They showered incandescent debris over the fragments and shards that were all that remained of his glass retorts and beakers, experimental chambers, even the giant burning glass that had first revealed oxygen to the world. Everything was destroyed.

During the course of this long night, Priestley had been planning his next sermon, to be preached among the ruins of the meeting-house using the text 'Father forgive them, for they know not what they do'. But when he felt the force of the mob's anger, he realised that all hope of reasoned arguments was gone. Harried constantly by fresh news of impending danger, he fled from this house to that, to London, and eventually to exile in America. He was safe there, and so was his family. But he was more than sixty years old and much of his life's work lay behind him, in ashes.

In France, Lavoisier had problems of his own. He had no particular reason to fear the revolution; indeed he had hailed it. Though extremely wealthy, he was no aristocrat, and he had long bemoaned the foolish way his country had been run, for the sole benefit of a privileged, and in his opinion often worthless, hereditary elite. The Ferme Générale's contract for tax collection had been cancelled by the National Assembly a few months earlier, but Lavoisier had already made his fortune and had neither the need nor the desire to continue his work there. Instead, his problems – at least at first – stemmed more from the energy that he felt obliged to pour into attempts at social reform.

As one of the most highly educated and progressive minds in the country, Lavoisier felt bound to devote virtually all his time to public service. He became chief financial adviser to the government, introducing a highly efficient system of bookkeeping into what had been the murky fog of national finance. He drew up detailed reports on the agricultural and industrial prospects of the country and demonstrated, in what one contemporary called a '*calcul très patriotique*' ('highly patriotic calculation'), that the nobles had composed barely 3 per cent of the population. With all these activities, no time remained for him to pursue his many scientific ideas, and he much regretted his own beloved, and now neglected, laboratory.

But worse trouble lay ahead for Lavoisier in the form of an old enemy, Jean Paul Marat. Marat's life story was an unfortunate one, full of desires he could never quite fulfil. Employed as a medical officer by an infamous royal aristocrat, the Comte d'Artois, he had plenty of opportunities to witness the privileges of wealth firsthand, but remained frustratingly unable to experience them for himself. The same went for science. He desperately wanted to make his name as a man of science, and once, years earlier, had presented a treatise to the French Royal Academy. In it, he declared that a candle goes out in an enclosed space because the air becomes dilated by heat and eventually smothers the flame. Lavoisier, a leading light of the Royal Academy, had been contemptuous of this work. It was not only wrong; worse than that, it was sloppy. Lavoisier, who required precision in all things, ensured that Marat's treatise was rejected, and he himself made sure that Marat was prevented from claiming the approbation of the academy when he wrote up his work.

Marat never forgot. And now, in the early years of the revolution, when little had yet changed in the lives of the poor people of Paris, Marat had become the spokesman for the mob. Finding himself at last in a position of power, he saw his chance for vengeance. He began to denounce this Lavoisier, 'son of a land-grabber'. In particular, he focused on one of Lavoisier's least popular achievements. In his position as a Fermier Général, Lavoisier had had a wall built around Paris. Typical of his mindset and reminiscent of his scientific experiments with air, it was a superbly efficient and accurate way to seal off the city and then catalogue every item that entered or left, to calculate the appropriate taxes. Marat's criticism of this was brilliant in its irony. He declared that this wall, built by the man who had given the world oxygen, had blocked the city's supply of air.

Lavoisier failed to see the danger. He made no effort to escape from Paris until the heat died down, nor did he seek to defend himself from Marat's foolish accusations. The world had always been good to him. Rational and scientific arguments had always prevailed, and Lavoisier saw no reason why that should change. But, like Priestley, Lavoisier was relying on reason at a time when reason had temporarily lost its grip. For the revolution had turned on itself. In this new time of Terror, a whisper was all it took, and Lavoisier found himself abruptly imprisoned, along with many of the other former Fermiers Généraux, without even knowing the charge.

The trial took place on 8 May 1794. Acts of accusation had arrived the night before, one for each prisoner, but candles were forbidden and it was too dark in the prison to read what later proved to be ludicrously trumped-up charges. The counsel appointed to defend Lavoisier didn't show up for the trial, and it probably would have made little difference if he had. The judge, one Pierre-André Coffinhal, was already famous for his barbarity and his love of playing to the crowd. After condemning a former fencing master to death he is said to have declared, 'Well, old cock, parry that thrust if you can.' He glibly refused to allow Lavoisier's carefully penned statements, or those of his friends and supporters, to be read before the court. He encouraged the jury to laugh uproariously at anything the accused tried to say. Several of the former Fermiers were unexpectedly whisked away partway through the proceedings; a word spoken somewhere in the right ear by the right person had earned them a last-minute reprieve. No such word came for Lavoisier. The bust of Marat stared down at him throughout the trial. When the verdict came, it was a formality. He and the rest of the remaining Fermiers had been found guilty of the capital crime of plotting against the Republic. A final plea for a two-week stay of execution to enable Lavoisier to complete some scientific work that would be of great value to mankind was dismissed by the judge in the now-famous words: 'The Republic has no need of savants. Justice must take its course.'

There was no more time for arguments. Lavoisier was taken immediately into an anteroom, where his hands were tied behind his back and his hair cut off at the nape of his neck. Next he was squeezed into a tumbrel with the other condemned prisoners, to make the short journey to the Place de la République. Lavoisier was the fourth to be executed, immediately after his father-in-law. The whole operation took about a minute: mount the steps to the scaffold, lay your head on the block, listen as the blade rattles back upward, ready to begin its descent. As the blade fell, Lavoisier took his final breath of the vital air that would make him famous. 'It took them only an instant to cut off that head,' said a friend and fellow thinker, astronomer Joseph-Louis Lagrange, when he heard the news. 'And a hundred years may not produce another like it.'

Oxygen is the most active ingredient in the air, but during their experiments Priestley and Lavoisier had also inadvertently isolated the other major ingredient of our atmosphere: nitrogen. This was the gas left behind when all the oxy-gene had been used up, the diluter that makes up some four-fifths of the atmosphere.[19] Nitrogen, it turns out, plays several important roles in the sustenance of life on Earth. It is one of the building blocks of the proteins in our bodies, which is why it is essential to eat certain vegetables that can 'fix' nitrogen directly from the air. But the other function is just as important.

Because Priestley was right, that if our atmosphere contained nothing but oxygen we would all be 'living out too fast'. In fact, much of the planet would burst spontaneously into flame. Dull, diluting nitrogen rescues us from having too much of a good thing, and for that we should all be grateful. As Priestley put it, 'A moralist . . . may say, that the air which nature has provided for us is as good as we deserve.'

These two elements – oxygen and nitrogen – make up the vast majority of the air that we breathe. But another substance in air is just as important for our survival. We need oxygen to burn our fuel, but the fuel itself comes from somewhere else. The source is another gas, present in the atmosphere in such tiny quantities that for many years it was presumed to be insignificant. And yet it is responsible for every scrap of food on Earth.

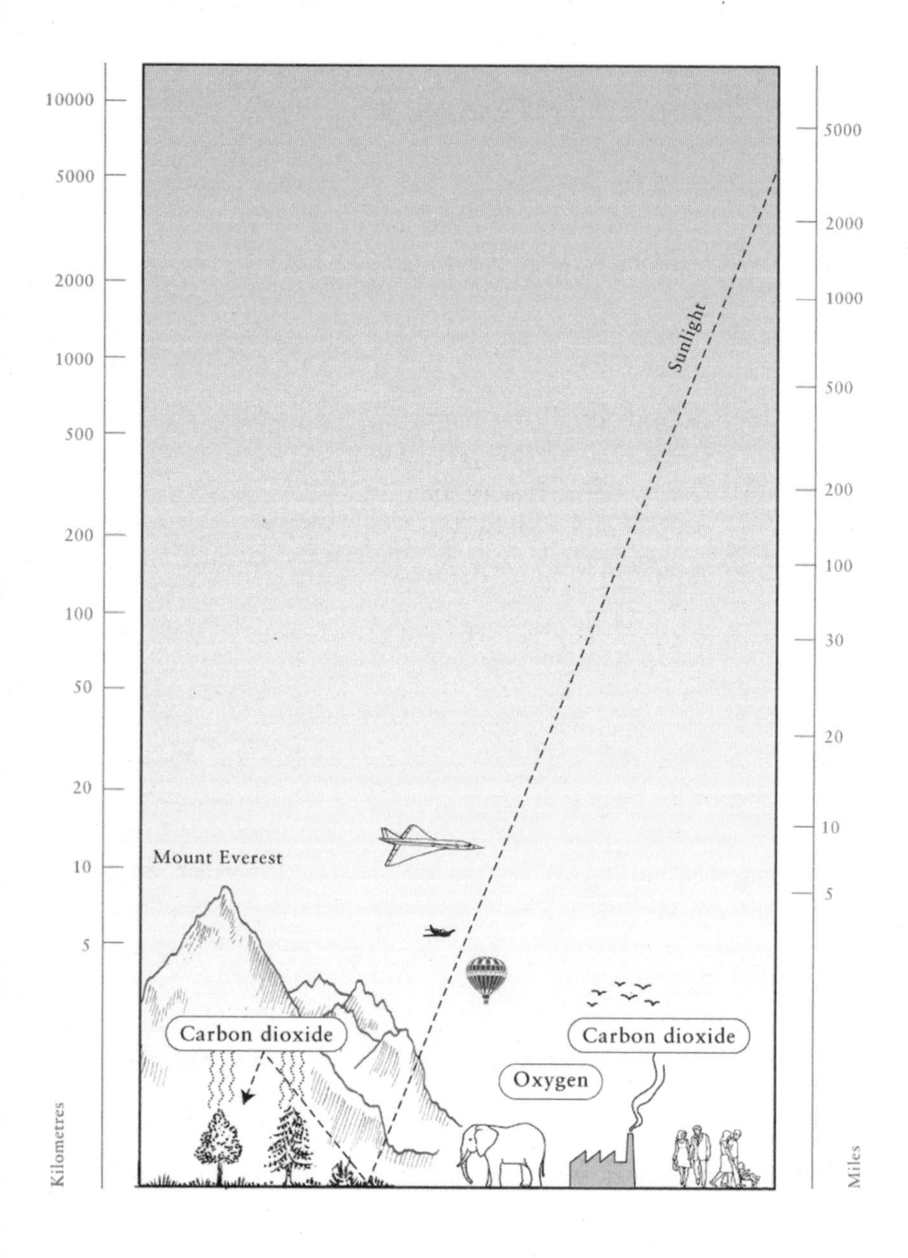

10000
5000
2000
1000
500
200
100
50
20
10
5
Kilometres
5000
2000
1000
500
200
100
30
20
10
5
Miles
Sunlight
Mount Everest
Carbon dioxide
Carbon dioxide
Oxygen

## *Chapter 3*

# FOOD AND WARMTH

The gas that feeds us all was discovered in the early eighteenth century, several decades before Priestley and Lavoisier would find its counterpart, oxygen. Its true identity was revealed by a gentle Scottish genius who created the gas twice, the first time idly, and almost by accident.

**January 1754**
**Edinburgh**

> I fully intended to have wrote last post, but really I happened to be intent upon something else at the proper time, and forgot it. It was indeed an experiment that I was trying that amused me, in which I had mixed together some chalk and vitriolic acid at the bottom of a large cylindrical glass; the strong effervescence produced an air or vapour, which, flowing out at the top of the glass, extinguished a candle that stood close to it; and a piece of burning paper, immersed in it, was put out as effectually as if it had been dipped in water: yet the smell of it was not disagreeable . . .

When Joseph Black wrote this letter to his former tutor, he had no idea how important his strange new 'air or vapour' would turn out to be. He was tinkering, amusing himself in between the more important matters of preparing his thesis and studying how better to cure his patients of their various ailments. For Black was training to be a medical doctor, and he took his profession seriously.

Everybody liked Black, and it was said of him that he never lost a friend. Sometimes he was too nice. He once placed all his funds in a financial house that then got into difficulties. Black realised there was a problem more than a year before the institution went broke but refrained from withdrawing his money for fear of causing embarrassment, and he ended up losing three-quarters of his savings.[1]

He was quietly confident, approachable, kind and almost impossibly curious. He loved experimenting, not just to find new medicines but also to see how the world really worked. Perhaps uniquely among academics then and now, Black truly had no desire for glory. Though he performed a huge number of experiments throughout his life, he hardly published any of them. He didn't want to be first, nor did he want to be famous; he simply wanted to *know*.

Black also enjoyed teaching. Later in life, when he became Professor of Anatomy at Glasgow University, most of his efforts went into preparing lectures that were extraordinarily popular with the students. He wasn't flashy, but gently enthusiastic and softly spoken, and his audiences remained so respectfully silent that his low voice could be heard from the back row. Black was above all steady; he could lift a beaker of vitriolic acid high into the air and pour it safely into a thin glass tube. Indeed, when he performed any of his demonstration experiments with acids and powders and colours and flames, his hand never shook.

Black never married, though he was a favourite of the ladies of Edinburgh. He dealt out his time and attention among them in carefully considered amounts, favouring the ones with suitably active minds. His closest friends were also lifelong bachelors, and, like Boyle, Black profited greatly from his interactions with them. They were dauntingly illustrious. Such was the concentration of talented minds in Scotland at the time that one famous historian in London remarked, 'I have always looked up with the most sincere respect towards the northern part of our island, whither taste and philosophy seem to have retired from the smoke and hurry of this immense capital.'[2]

In Edinburgh with Black were philosopher David Hume; Adam Smith, the father of modern economics, and James Hutton, who created the science of geology. While natural scientists in London were still stuck on the old favoured topic of the stars, those in the new industrial centres had begun on another tack. Like Galileo before them, they wanted to switch focus. Never mind the heavens, they were saying, what do we have *here*?

Black's famous friends were just as agreeable as he was. The four of them set up a weekly discussion meeting, called the 'Oyster Club', which was open to any inhabitants of Edinburgh, or indeed visitors, with interests in art or science. The talk was informal, and none of the founders was intimidating or aloof. As one commentator noted, the four friends were easily amused, just as ready to listen as to speak, and 'the sincerity of their friendship had never been darkened by the least shade of envy'.[3]

Black's life, however, was considerably darkened by the ill health that constantly dogged him. He was often frustrated by the slow pace at which he was forced to pursue his studies. Several days of intellectual effort at a stretch was enough to start him coughing up blood, and he sometimes excused his failure to reply to his father's letters by saying that he was 'too miserable'. In his later years he became steadily more frail, stretching out his life through careful exercise and an increasingly dismal diet. When he finally died, it was while eating his usual slice or two of bread and a few prunes, washed down with milk diluted with water. A servant found him already dead, still balancing the cup of milk on his knee, 'as if', one friend later wrote, 'an experiment had been required to show to his friends the facility with which he departed'. He hadn't spilled a drop.

When Black began his work on airs, it was almost by accident. Ever the medical man, he was trying to find a cure for an excruciatingly painful illness that plagued the people of the seventeenth century just as it continues to plague us today: bladder stones. While we now have fairly humane cures, in the seventeenth century, without the benefits of sterile instruments or anaesthetics, surgery was a potentially fatal endeavour. More indirect treatments involved injecting the bladder with caustic substances that would certainly help to dissolve the stone but would dissolve a lot more besides, and they often ended up being more painful and debilitating than the disease they were trying to cure.

In an effort to avoid these alternatives, sufferers turned to increasingly bizarre concoctions. The British prime minister, Sir Robert Walpole, who made his discomfort from the stone very widely known, ensured that a certain Mrs Joanna Stephens received five thousand pounds for revealing the secret of a recipe that he believed had helped him. In the *London Gazette* on 19 June 1739, Mrs Stephens reported that her brew consisted of:

> . . . a Powder, a Decoction, and Pills. The powder consists of Egg-shells and Snails, both calcined. The decoction is made by boiling some Herbs (together with a Ball, which consists of Soap, Swines-Cresses, burnt to a Blackness, and Honey) in water. The Pills consist of Snails calcined, Wild Carrot seeds, Burdock seeds, Ashen Keys, Hips and Hawes, all burnt to a Blackness, Soap and Honey.[4]

Black had little patience with such mystic brews and wanted to find a cure that was more scientifically based. He decided to start with a powder called *magnesia alba*, made from Epsom salts, which he already knew to be both mildly caustic and useful in medicine. He had prescribed it, for instance, to 'an active woman of rather full habit of body, and it purged her ten times', and concluded that 'this salt, though mild to the taste, seems yet to surpass other purgatives'.[5]

The idea was to try to induce *magnesia alba* to produce a new product that was caustic enough to dissolve the stone but mild enough to cause less discomfort than the usual treatments. He decided to try heating it and then mixing the result with water, which was the standard way of making caustic medicines. He placed an ounce of *magnesia alba* in a crucible and blasted it with enough heat to melt copper. Much to his surprise, he discovered that this drove off every shred of causticity. The resulting white powder was milder than ever, had no effect on the water he tried to mix it with, and even refused to fizz when added to an acid. It would never be a cure for the stone.

Black, ever careful, weighed his sample after the experiment and discovered that he had finished with 'three drachms one scruple', which was just five-twelfths of the original weight. He was mystified. There may have been a little water in his sample, but not nearly enough to account for such a drastic loss in weight. Where had the rest of the *magnesia alba* gone?

Putting aside his disappointment at this failure to produce a cure for the stone, Black decided to try to find out. Since the *magnesia alba* had clearly not lost enough water to account for its change in weight, the only other alternative was air. And this reminded Black of the work of a clergyman who, nearly thirty years earlier, had published a book about his strange experiments with vegetables.

Stephen Hales had a straightforward, if simple-minded, approach to his profession. His style at the pulpit was all fire, brimstone and damnation. To be sure, he often preached the Christian duties of charity and generosity towards the poor, but he was also constantly watching for any evidence of 'disorderly' or 'loose' behaviour among his parishioners. He campaigned against swearing and, though he was partial to wine himself and had no problems with cider and ale for the lower orders, he warned sternly against the drinking of spirits such as gin and brandy. Though much of his detestation of spirits came from a conviction of their harmful effects on the body, he also disliked the inclination of those drinking them towards loose morals, warning rather poetically of the 'bewitching of Naughtiness in these fiery liquors'. His ideas for penance were also old-fashioned. Some unfortunate parishioners found guilty of fornication were made to stand barefoot outside the church, wearing a white sheet and holding a white rod, until just before the litany, at which point they were brought inside to hear the sermon and be prayed for.

Though Hales's Sundays were spent haranguing his flock, much of the rest of the week was devoted to his other passion: science. While waiting for the job at his parish to become available, he had spent nearly thirteen years at Cambridge University, where the great Sir Isaac Newton was still in residence, and his interest had been aroused. Now, at his parish in Teddington near London, he spent large amounts of time poking things, prodding them and cutting them up like a curious schoolboy. He saw no conflict between his two favourite activities, religion and science. Instead, like Boyle, he decided that the more he discovered about the workings of the world, the more thoroughly he believed. 'What a multiplicity, variety, beauty, usefulness, and subservience to each other, may we with pleasure observe, in contemplating the Works of Creation,' he declared.

In fact, the only conflict came from his experiments on animals. For instance, in comparing the circulation of blood in the human body with the circulation of sap in trees, he performed some gruesome experiments on luckless dogs, horses and deer, until he decided that it ill-behoved a man of God to continue. He wrote to a fellow clergyman that, since further experimentation along these lines would require the death of several hundred animals, 'I do not think it proper for one of our profession to engage any further in it.'[6]

Instead of cutting up God's fellow creatures, Hales decided to try heating every kind of natural, but inanimate, stuff he could find. He tried hog's blood, deer's horn, peas, tobacco, oil of cloves, beeswax, and even the stones from a human gall-bladder. And this is the point where his random experiments became suddenly very important. Because Hales found that when you heated these substances, every single one gave off air.

For scientists at the time, this was astonishing, like conjuring up a genie from Aladdin's lamp. Obviously liquids, water for instance, could turn into vapour when they boiled. But how could something as insubstantial as air be trapped inside a solid? What's more, there was so much of it.

In his book *Vegetable Staticks*, published in 1727, Hales reported that:

> There arose from a piece of heart of oak, 216 times its bulk of air. Now 216 cubick inches of air, compressed into the space of one cubick inch, would, if it continued there in an elastick state, press against . . . the six sides of the cube with a force equal to 19860 pounds, a force sufficient to rend the Oak with a vast explosion.[7]

Since Hales was no fool, and he had noticed that oak trees do not generally explode without warning, he decided that the air he had released must somehow previously have been fixed in place. Hales imagined that his 'fixed air' was made of particles that repelled each other mightily. In some circumstances, he believed these particles could become bound inside his solid objects and in others, liberated again.

However, all Hales cared about was how air became fixed, and how it subsequently recovered its bounce. He had no idea what was truly happening when gas flooded so unexpectedly out of a solid, nor did he realise that individual 'airs' might have different properties.

Steady, thoughtful Joseph Black was much better placed than the rambunctious Hales to figure this out. Inspired by Hales's work, he wondered whether his *magnesia alba* had been transformed by losing some quantity of fixed air. That, at least, would explain why it lost so much weight. Moreover, rather than assuming that every air was the same, just with more or less bounce, Black suspected that Hales's fixed air might have properties of its own, ones that could be quite different from those of ordinary, common air. Perhaps it even had enough individual characteristics to explain why the *magnesia alba* had lost its causticity and turned so mild after it vanished.

Black didn't manage to catch any gas in the act of escaping from *magnesia alba*, but had more success with one of its caustic cousins: marble. He heated a cubic inch of the stuff, and – sure enough – produced a huge amount of fixed air, enough to fill a vessel holding six gallons.

Now that he had some samples of fixed air to work with, Black determined to find out whether it was truly different in character from common air. The experiment he designed to test this was rather complicated, but also ingenious. Black knew that lime water (which is just lime, or calcium, dissolved in water) had an affinity for fixed air. He decided that the lime in lime water must be soaking up the air, the exact opposite reaction to the way that *magnesia alba* and marble had released it. He also knew that water always has a certain amount of common air dissolved in it. That is why fish can breathe underwater, and why tiny bubbles form long before a pot of water comes close to boiling.

So he wondered what happened to the common air dissolved in lime water. If common air were just the same stuff as fixed air, any common air in lime water would be sucked up by the lime, leaving none behind in the water. Black realised that all he had to do was check how much common air was dissolved in equal quantities of ordinary water and lime water. If equal amounts came from both, the air the lime sucked up must be fundamentally different. And that would mean that his new air really was special.

To put his idea into practice, Black needed an air pump. But the only one available in Edinburgh was frustratingly out of action, and its slow, surly technician was impervious to Black's pleasant requests that he speed up his attempts to fix it. Exasperated, Black wrote to his former tutor in Glasgow, begging him to use the air pump there and explaining with great precision exactly how the lime water should be made and treated. His tutor quickly arranged for the experiment to go ahead. Word came back. Four ounces each of lime water and ordinary water had been placed under the receiver of the new Glasgow air pump. As the pump sucked, air bubbled up out of each of the two vials. Each released almost exactly the same amount.

Black was delighted. 'From this it is evident,' he wrote in his thesis, 'that the air which quicklime attracts, is of a different kind from that which is mixed with water . . . Quicklime does not attract air when in its most ordinary form, but is capable of being joined to one particular species only, which is dispersed throughout the atmosphere.'[8] In honour of Hales, Black decided to call this extraordinary new species 'fixed air'. We know it now as carbon dioxide.

In the history of science, this apparently innocent moment was in fact extraordinarily profound. For this was the first time that anyone had shown there was more than one kind of gas. Because of this discovery, Black would be known as the father of modern chemistry. Lavoisier, Priestley and their contemporaries all regarded themselves as Black's disciples. Lavoisier, usually reticent about offering credit to others, even wrote to Black saying how much he admired his work.

But more important for our story is the nature of the gas he had found. Ever curious, Black decided to abandon his work on bladder stones for a while and find out how his new fixed air behaved. He remembered the old experiment that he had described to his tutor back in January. Sure enough, adding acid to chalk produced the same fixed air that had flooded out of marble. Black also found that he could make it by simply burning charcoal in ordinary air. And, as before, though the fixed air smelled 'not disagreeable', it snuffed out candles and animals could not breathe it and live.

Black also noticed that fixed air is a product of distillation and that it appears in our breath when we exhale. He was, however, baffled by what it could be doing in our bodies in the first place. 'It is not to be doubted,' he wrote, 'indeed that this air, extensively united with every part of our body, serves many great uses, nor is it to be supposed that its absence could be borne without inconveniences: but we do not seem to know what its use is, or what are the inconveniences that would result from its absence.'[9]

The 'inconveniences', it turns out, are that without it we and most other living things on Earth would starve to death.

Black never knew the vital role that his discovery, carbon dioxide, plays in our lives, but those who came after him quickly began to recognise its importance. In Lavoisier's experiments with respiration, he realised that the more oxygen a person or animal consumed by breathing, the more 'fixed air' they produced. He deduced that we burn our carbon-based food in much the same way that a candle burns its carbon-based wax, and for the same reason: to release energy. And burning carbon-based substances in oxygen produces – what else but carbon dioxide.

Priestley, meanwhile, spotted that the interplay between fixed air and oxygen was somehow related to plants. He knew that a mouse in an enclosed chamber would eventually be unable to breathe, but he discovered that placing a plant in the same chamber kept the air from getting noxious indefinitely. The plant and mouse seemed to work in contented co-operation to keep the air fresh.

This is not merely a curiosity. Subsequent scientists have discovered that it's the fundamental basis for life as we know it on Earth. For the existence of carbon dioxide, and its relationship with oxygen, is the foundation for a pact between plants and animals the world over.

We animals take in oxygen to burn our food and throw out carbon dioxide as a waste product. Plants work the other way round. They take in carbon dioxide to *make* food and produce oxygen as their waste product. (Plants also need to breathe, to release energy from the food they make. They use up about a quarter of the oxygen they produce, but the rest they leave for us.) So we have a deal that keeps us all alive: plants soak up our leavings and we soak up theirs. Air is the living, breathing medium for this eternal interchange.

The plant's side of this bargain is the basis for all food production on Earth. The first hint that this might be so had come in the mid-seventeenth century, when a Dutch alchemist named Jan Baptista van Helmont performed a curious experiment. He had begun to wonder what plants are made of, or more particularly where the stuff that makes a plant comes from. So he took a large pot and put in it two hundred pounds of earth that he had carefully dried in a furnace. In this pot he planted a young willow sapling, weighing five pounds. And over the top of the pot, so that no extra dust could enter from the air, he fitted a metal plate full of holes around the sapling's trunk. Van Helmont was a persistent fellow. He pursued his experiment for a full five years, watering, watching, and waiting. In the end, he had a towering willow tree that weighed '169 pounds and about three ounces.'

So where had the tree come from? The first thing to test was the earth in the pot. Van Helmont removed the earth, dried it and weighed it. It had lost a mere two ounces.

This might not seem so surprising. After all, anyone who has ever owned a house plant knows that it will grow happily without your adding new soil to the pot. But in that case, what had made the willow tree's branches, trunk and leaves?

Van Helmont guessed wrongly. The only thing he had added to the pot was water, so he blithely declared that water had to be the source. (He wasn't brilliantly logical in his deductions in other matters, either. Among other odd beliefs, he was convinced that living things could arise spontaneously out of the strangest ingredients. He even published a recipe for making mice out of dirty underwear and wheat: 'For if you press a piece of underwear soiled with sweat together with some wheat in an open mouth jar, after about 21 days the odour changes and the ferment coming out of the underwear and penetrating through the husks of the wheat, changes the wheat into mice.')

The problem in this case was that he hadn't even noticed that the tree was surrounded by something else that was a superb source of raw material for making plants: thin air. The source of every ounce of the solid roots, trunk, branches and leaves of Van Helmont's willow tree was the carbon dioxide in the air around it.[10] When plants soak up carbon dioxide, they take air and turn it into the food that eventually finds its way into our stomachs.

Plants do this in a complex series of internal reactions, but the overall result is a simple one. They use the sun's energy to break apart carbon dioxide and turn it into the carbon-based molecules that make up our food. The scale of this activity is staggering. Every year, green plants convert carbon dioxide into more than 100,000 million tons of plant material.[11] To do this, plants use up 300 trillion calories of energy from the sun, which is thirty times the energy consumption of all the machines on Earth. Even the animals we eat gain their protein and fat from plant food. Carbon dioxide in our atmosphere is the fundamental foodstuff for every plant, animal and human on the planet.

Trees and plants take their nutrients from our ocean of air in the same way that waving fronds of seaweed do from seawater. And when we breathe, we simply recombine the food they made with the oxygen they produced to start the process all over again. The balance isn't perfect, and that turns out to be a good thing. The only reason we have oxygen to breathe in the atmosphere today is that plants keep hold of a certain percentage of the stuff they make and prevent us animals from eating it, breathing it and turning it back into carbon dioxide. The fraction is small, just 0.01 per cent of the stuff that plants make, but that also means the same percentage of the oxygen they make also remains free to float up into the sky. Over billions of years, this has built up into the atmosphere we need to live.

Some researchers even see the pact between plants and animals as being more like a battle.[12] At certain times in the past, plants have had the upper hand. For instance, a little over 400 million years ago, plants discovered how to make lignin, the hard stuff that turns into the woody parts of trees. Nothing in the animal kingdom knew how to digest this strange new material, so it remained untouched and unrespired – and a little less carbon dioxide made it into the atmosphere.

Then came the two champions of the animal kingdom: termites and dinosaurs (the vegetarian sort). Both learned how to digest lignin, and carbon dioxide levels rose again. Until, that is, the extinction of the dinosaurs, when plants learned how to make vast grasslands and the balance swung again.

This mattered for much more than plant pride. It turns out that interfering with the amount of carbon dioxide in the atmosphere can have serious consequences. As well as providing our food, carbon dioxide plays another role, which is every bit as crucial in shaping our planet for life.

The man who discovered this was John Tyndall, an exuberant Irish physicist who was a professor at London's ultrafashionable Royal Institution in the mid-nineteenth century.

The Royal Institution was the perfect place for someone like Tyndall; he could perform his research in the basement laboratories and then talk about science in the famous lecture theatre aboveground. Science had become one of the hottest entertainments in town. Lectures at the institution were so popular that, to cope with the crush of carriages, Albemarle Street became Britain's first one-way street. And it wasn't only scientists who were crowding on to the Royal Institution's uncomfortable wooden benches. There were poets and politicians, intellectuals and aristocrats, in fact most of London's *beau monde.*

Tyndall loved lecturing. Perhaps because he had come to research late, beginning his higher education only in his late twenties, he couldn't wait to pass on his findings. He was less concerned about education than about sharing his own wonder.

He choreographed his lectures as for a West End show and worried endlessly about how to ensure their success. One day when he was preparing a lecture, Tyndall knocked an instrument off the table but managed to vault over and catch it before it reached the ground. He was so delighted with the effect that he practised it for hours. When he 'accidentally' repeated the feat that evening, he brought the house down.

These efforts paid off. When word went out that Tyndall was lecturing, the house was always packed. And not just at the Royal Institution. Tyndall's lectures to illiterate working men at the Royal School of Mines attracted audiences of six hundred or more. One contemporary commentator wrote: 'Professor Tyndall has never for an instant looked upon the masses as entitled to only second-rate knowledge. They have had it of the highest and purest which it was in his means to supply.' And during a lecture tour in America, the *New York Daily Tribune* said of him:

> There is no such thing as doing justice by description to Professor Tyndall's manner. It is so pleasant, so colloquial, so free of arrogance, so full of personal enthusiasm as if the wonders he displayed were as new to him as to the rest of us. He makes science easy, coaxing his audience over the hard places by promises of untold beauties to come. In short he is the very beau-ideal of a scientific lecturer.[13]

Tyndall was impulsive, passionate and sincere. He had a large nose that jutted out to a point, with two deeply grooved lines running from either side in a graceful arc down to the edges of his mouth. In later years he sported an impressive white beard, in true Victorian style, which sprouted around his chin and neck, though he kept his face clean-shaven. He could be intense and sometimes self-righteous but he also had his playful side, and children loved him. Practical jokes were more his style than verbal witticisms, though, and he was apt to greet any wordplay with what a friend, evolutionist Thomas Huxley, once described as 'blank, if benevolent, perplexity'.

Along with Huxley and seven other friends of a scientific bent, Tyndall was a founding member of a discussion group that became famous as the 'X Club', so called because, even after many hours of disputation, nobody could agree on a better name. The founders also expended much time on discussing the possible addition of new members, until this grew so tiresome that they agreed that no proposition of that kind should be entertained unless the name of the new member suggested contained all the consonants absent from the names of the old ones. 'In the lack of Slavonic friends,' Huxley said later, 'this decision put an end to the possibility of increase.' Tyndall's membership of this club, coupled with his often obsessive leanings, earned him the nickname 'Xccentric'.

Some of Tyndall's poet friends complained that learning science could deaden one's appreciation of nature, but Tyndall himself was exasperated by this attitude. For him, the better he understood the world, the more wonderful he found it, and his skill at explaining carried many others along with him. He said that science *required* imagination. (In fact a phrase he coined, 'the scientific use of the imagination', was later quoted by Sherlock Holmes in *The Hound of the Baskervilles.*)

In particular, Tyndall was fascinated by the happenings of the invisible world of atoms and molecules. At the time there were no microscopes capable of capturing the motions of these minuscule entities in action; the only way to study them was to combine logical thought with a vivid imagination. Tyndall had both talents in abundance. Huxley said of him, 'In dealing with physical problems, I really think that he, in a manner, saw the atoms and molecules, and felt their pushes and pulls.' Tyndall thought so, too. At the end of a lecture about radiation, he said, 'It is thought by some that natural science has a deadening influence on the imagination . . . But the . . . study of natural science goes hand in hand with the culture of the imagination. Throughout the greater part of this discourse . . . we have been picturing atoms and molecules and vibrations and waves which eye has never seen nor ear heard, and which can only be discerned by the exercise of imagination.'[14]

This capacity to picture and understand the invisible was the perfect background for studying the behaviour of air. But at first Tyndall paid little attention to the atmosphere. He was more interested in the studies of magnetism and the compression of crystals. However, this led to an interest in the movement of glaciers, and it was during field trips to the Alps to study these phenomena that Tyndall's interest in the atmosphere was first kindled.

Tyndall loved the mountains. He was sure-footed, a strong and daring climber. To follow his scientific nose, Tyndall would cheerfully hack his way up ice cliffs, dodging falling rocks, or plough his way through fields of crevasses. Once, making his way in the name of science through the seracs of the Glacier du Géant, he felt truly terrified. But afterwards he described the scene with relish:

> Wherever we turned, peril stared us in the face . . . Once or twice, while standing on the summit of a peak of ice, and looking at the pits and chasms beneath me . . . I experienced an incipient flush of terror. But this was immediately drowned in action. Indeed the case was so bad, the necessity for exertion so paramount, that the will acquired an energy almost desperate, and crushed all terrors in the bud.

During his trips to Switzerland, Tyndall became entranced by the alpine skies. 'The shiftings of the atmosphere were wonderful,' he wrote after one day out on the mountains, and 'half the interest of the Alps depends on the caprices of the air', after another. He even began to feel connected to the air in a way that he had never experienced before. 'In effect,' he said, 'we live *in* the sky rather than under it.'

Once his attention was caught by air, Tyndall was immediately gripped by the urge to understand it. Trips to the mountains were always undertaken for scientific purposes. After all, how can you appreciate the landscape if you don't try to make sense of it? This view was not always shared by the less scientifically minded members of the Alpine Club. One year, at the club's winter dinner, the speaker gave a sarcastic sideswipe at Tyndall's scientific obsessiveness. He was describing a mock ascent of a mountain, and finished by saying:

> 'And what philosophical observations did you make?' will be the enquiry of one of those fanatics who, by a reasoning process for me utterly inscrutable, have somehow irrevocably associated Alpine travel with science. To them I answer, that the temperature was approximately (I had no thermometer) 212 degrees Fahrenheit below freezing point. As for ozone, if any existed in the atmosphere, it was a greater fool than I take it for.

Tyndall never took his science lightly. Deeply offended, he instantly resigned from the club.[15]

Tyndall hoped that studying the atmosphere might help him explain a conundrum furnished by the mountains themselves. His beloved Alps were full of evidence that at some point in history there had been an 'ice age'. Valleys had been scoured out by glaciers that had long since vanished, rocks had been transported by ancient ice far beyond their places of origin, and jumbled piles of rubble and moraines, delineated where existing glaciers had once dramatically extended their reach. How could the world have once been so cold, and what made it warm up again? Tyndall wondered if slight changes in the atmosphere might be the answer.

In particular, Tyndall suspected that the atmosphere might act as a blanket around the world, sometimes warming and sometimes cooling as the components slightly shifted their relative proportions. He thought this because of an effect spotted a few decades earlier by French scientist Joseph Fourier. Fourier had noticed that Earth should, by rights, be much colder than it actually is. We tend to think that Earth lies in the perfect position for habitability. Of our two nearest neighbours, Venus is too close to the sun and too hot to sustain life, and Mars is too far from the sun and too cold. Yet Earth is 'just right', the perfect distance for running water, balmy breezes and a comfortable, temperate planet. However, Fourier realised that we're actually a little too far from the sun to survive without help.

When sunlight arrives to warm Earth, the energy it provides doesn't simply stay put. Like a central-heating radiator, the warm planet starts pouring heat energy back out into space. The balance between these two effects sets the planet's thermostat. And when he calculated the difference between the heat energy arriving and leaving in this way, Fourier was perturbed by his findings. Earth should be perpetually frozen.

Fourier had guessed that something in the air might help to trap extra heat on the planet's surface, and explain why we are so comfortably off, but he didn't know what. Thinking about Fourier's earlier work, Tyndall decided that he agreed. And if he could find this mysterious warming component, he might begin to understand how our climate could have been different in the past.

So, in the summer of 1859, Tyndall set about constructing an artificial sky in the basement of the Royal Institution. It was a splendid piece of Victorian scientific equipment, a long tube filled with gases and surrounded by sources of heat and light, and pipes that looked like the flailing tentacles of an octopus.

Tyndall enjoyed playing with his mini-atmosphere. He shone white light through it and discovered that tiny particles in the air scattered blue light much more than all the other colours of the rainbow. This, he surmised, could explain why the sky is blue.[16] A similar effect happens in the oceans, with scattering from tiny bits of mud. Illustrating this point in a lecture, Tyndall said,

'And thus the blue eyes so admired among the ladies of my audience owe their charm essentially to muddiness.' You can see this 'Tyndall effect' for yourself if you're ever out in a car on a foggy night. Scattering from the particles of water in the fog will turn the light from your headlights a fetching shade of blue.

But what Tyndall really wanted to know was how the atmosphere retains more heat than it should by rights. He considered both sides of the heating equation. First, the ordinary visible sunlight that comes in to heat Earth. Obviously this must slip through the sky unimpeded or it couldn't arrive at the surface; the sky would be permanently dark and we wouldn't see the sun, moon or stars. However, perhaps the answer lay on the other side of the heating balance, the part where Earth radiates energy back out into space.

Everything that is warmer than its surroundings radiates heat. You do it, I do it, and so does every warm-blooded animal. But we don't see each other permanently glowing because the light we give off is invisible. There's much more to light than the ordinary visible rainbow. Just as there are sounds too high- and low-pitched for us to hear them, so some 'pitches' of light evade our eyes. In this case, the invisible light is called infrared. It lies just over the edge of the red part of the rainbow, its frequency too low for us to see. Infrared light is the means by which remote controls communicate with televisions and stereos, and how 'night-vision' goggles can show people moving around with ghostly glows even in the pitch black. It's also how our planet pours its heat back into space.

Tyndall knew all about infrared light. He decided to investigate whether the atmosphere traps infrared light on its way back out into space, and so keeps our planet warm. But what gases should he include in his artificial atmosphere? By now, 150 years after the pioneering experiments of Joseph Black, science had progressed mightily. Everybody knew that the atmosphere was made up of many different gases, but that most of them were present only as tiny whiffs. Since by far the bulk of the air consists of nitrogen and oxygen, Tyndall started with these. But try as he might, he couldn't get his air to take up infrared light. The light simply slipped through, taking its heat with it.

And then one day, without much hope that it would make any difference, Tyndall decided to try another component of the atmosphere: carbon dioxide. It seemed a long shot. After all, air contains nearly 79 per cent nitrogen, 20 per cent oxygen, and barely 0.004 per cent carbon dioxide. Such an insignificant gas could hardly explain something so momentous.

Nonetheless, Tyndall shone his source of heat – a copper cube filled with boiling water – on to one side of his model atmosphere and watched what happened. To his amazement, the needles of his instruments immediately began to lurch. Even in such tiny amounts, carbon dioxide turned out to be a monster absorber of infrared light.

Carbon dioxide absorbs infrared light so well because each individual molecule is relatively big and complex. Molecules soak up light energy because they want to vibrate like a tuning fork or tumble like an acrobat. And complex molecules have many more ways to do this than the more simple varieties. Brilliant, imaginative Tyndall realised this long before advanced technologies showed it to be true. He said, 'The *compound* molecule . . . must be capable of accepting and generating motion in a far greater degree than the single atom.' Oxygen ($O_2$) and nitrogen ($N_2$) are not single atoms – each is made up of two individual atoms of the same element. But they're still too simple to soak up infrared radiation – they don't have enough options for how to move. But carbon dioxide is a different matter. It's made up of one atom of carbon and two atoms of oxygen, and it can vibrate and spin with abandon. That's why it's such a good absorber of radiation, and also why a little carbon dioxide goes a very long way.

Tyndall discovered that water vapour is an even better absorber of infrared radiation. In fact, our atmosphere is full of infrared absorbers, including methane, ozone and the man-made chemicals that also bedevil the ozone layer. By far the biggest warming effect comes from water vapour, not because it's the most effective pound for pound – it isn't – but because there's so much of it in the sky. However, carbon dioxide is still a significant climate-driver, because even small changes in the gas can make big differences to the temperature. Since warmer air soaks up more water vapour from the ocean, the two gases, carbon dioxide and water, work together to wrap the planet in a comfort blanket of warmth that keeps us all alive.

This insight from Tyndall was the beginning of our understanding of the impact the famous 'greenhouse effect' has on Earth's climate. 'Greenhouse' in this case is actually a misnomer, since greenhouses work mainly by trapping the air inside them. The glass windows allow light to enter and warm the air, but they also prevent this newly warmed air from wafting away. The gases in our atmosphere don't work quite like this. Rather than keeping warm *air* in place, they catch the infrared radiation on its way from the surface out into space. They vibrate with the energy for a brief instant and then throw the energy back out like a fielder returning the baseball he's just caught. Since, unlike most baseball fielders, they throw out their energy wildly in random directions, some of it succeeds in escaping into space. But enough is hurled back down to Earth to keep our lifeblood – water – from freezing.

Tyndall's description of this effect was typically poetic. Without it, he said, 'The warmth of our fields and gardens would pour itself unrequited into space, and the sun would rise upon an island held fast in the iron grip of frost.'

To Tyndall and his contemporaries, carbon dioxide was not the menace we see it as today; indeed, it was a lifesaver. But he also realised that, because there's so little carbon dioxide in the atmosphere, even small changes in the past could account for dramatic swings in climate, such as the ice age that had left such a mark on the Alps. This, he said, might explain 'all the mutations of climate which the researches of geologists reveal'.

Though Tyndall didn't realise it, this was the first hint of carbon dioxide's downside. Yes, it is the crucial source of all our food, and yes, without it we would freeze to death. But, like oxygen, carbon dioxide has the potential to deliver too much of what would otherwise be a very good thing. The hero that protects us was also to be revealed as a villain, threatening us with a potentially deadly menace: global warming.

**1896**
**Stockholm, Sweden**

Svante Arrhenius was depressed. Now aged thirty-seven, he had just been through a messy divorce and had lost not only his wife, but also custody of their young son. The pouches under his eyes and the moustache that plunged sharply downward on either side of his mouth only served to emphasise his present misery. He urgently needed something to distract him. But what?

Arrhenius was a scientist. Mainly he worked on the chemistry of liquids that conduct electricity. In five years' time he would win a Nobel prize for his research. This would embarrass his thesis examination committee, who had labelled his work 'mediocre' and barely let it scrape by. But though he was fascinated by his usual subject, he was looking for something different to dabble in for a while. What he wanted most was a change.

That's when he happened on Tyndall's idea of the role that carbon dioxide might play in causing the ice ages. Arrhenius was intrigued by this notion and wanted to take it a bit further. Being a theorist, he decided to calculate exactly how much carbon dioxide Earth would have to lose to trigger an ice age.

This was going to be more complicated than it first seemed. For Arrhenius realised that he couldn't just stick to the direct cooling caused by less carbon dioxide in the atmosphere. There would be some important additional effects as well. In particular, he knew that cooler air is a less effective sponge – it soaks up less water from the oceans.

That matters because, as Tyndall had noticed, water vapour is a very effective greenhouse gas in its own right, and losing some of it would make the atmosphere colder still. In other words, a small change in carbon dioxide could make a big difference in climate. (This highlights an important aspect of the way that carbon dioxide levels can affect our climate. Many sceptics have pointed out that water vapour is the principal source of greenhouse warming in the atmosphere; in terms of absolute effect, carbon dioxide comes a very poor second. However, Arrhenius was right that changing the amount of carbon dioxide by only a small amount significantly changes the amount of water vapour, which boosts the overall impact. By this means, carbon dioxide punches considerably above its weight.)

Arrhenius realised that if he wanted a plausible answer, he would have to consider both the direct and indirect effects of carbon dioxide in tandem. This would entail long, tedious calculations. Perfect. It was just the sort of distraction he had been looking for. He picked up pencil and paper and settled down for several months' hard labour.

First, he imagined a world in which carbon dioxide was reduced by half. He then carefully calculated the amount of moisture in the air, and the amount of light energy entering and leaving Earth, for every zone of latitude. Eventually, he had an answer. It was crude, with many assumptions, but it was the first attempt to put numbers on the effect of changing carbon dioxide. Halving the $CO_2$ levels would reduce global temperatures by about 5 degrees Celsius. That, he thought, could be just enough to trigger an ice age.

Arrhenius was a theoretical chemist, not an atmospheric scientist. He had picked the amount of carbon dioxide to try almost at random, and he had no idea whether it was realistic. So he asked a colleague's advice. Arvid Högbom had already come up with numbers for how much carbon dioxide appears naturally from volcanoes and how much disappears into Earth's rocks and oceans. It should certainly be possible to reduce levels of carbon dioxide, he said, if some volcanoes dried up for a while, or something happened to stop the oceans from soaking it up. But while he was fiddling with the numbers, Högbom noticed something curious. Never mind lowering carbon dioxide levels; they were already being *raised* in a way that had nothing to do with volcanoes or oceans, or indeed any other natural process. To keep the factories of the Industrial Revolution running, humans were now burning coal on an unprecedented scale. In the process, they were producing tons of carbon dioxide. When he compared this with the natural sources, Högbom found that humans were producing carbon dioxide at the same rate.

Högbom wasn't particularly alarmed by this. After all, even in 1896, at what seemed like the height of the Industrial Revolution, a full year's worth of coal wouldn't increase carbon dioxide in the air by very much – perhaps only one part in a thousand. He had no concept – nobody did – of how dramatically the world's population would increase, and how much industrialisation would accelerate. But his results did set Arrhenius thinking.

He realised the heating process would be almost the exact mirror image of cooling. Just as cooler air holds less water vapour, so warmer air holds more. So more carbon dioxide would warm the air in its own right, and also encourage more water to evaporate from the oceans, which would warm the air even more. Once again, Arrhenius worked through his calculations. If, say, carbon dioxide were to double from its 1896 level, even though that would still be only a tiny fraction of the air as a whole, Arrhenius predicted that it would cause a large amount of warming, perhaps changing temperatures by as much as 5 degrees Celsius. Though that might not sound like much, raising global average temperatures everywhere in the world by this amount could make a huge difference to the overall climate. (Amazingly, it is also very close to the result that many computer models produce today, using sophisticated calculations and vastly improved knowledge of how the climate works. More than a hundred years ago, Arrhenius was on the right track.)

Arrhenius's finding caused a flicker of interest but not much concern. Assuming industry continued at more or less the same rate, doubling carbon dioxide levels would take thousands of years, and the calculation seemed like a curiosity rather than a cause for worry. And even if warming happened more quickly, this was the time when technology was almost universally considered to be a good thing. Who was to say that a warmer world wouldn't be a better one? Another scientist of the time, Walter Nernst, thought it would. He suggested deliberately heating up Earth a bit, by burning useless coal deposits.[17]

But even this kind of speculation died down when experiments appeared to show that Arrhenius had got his calculations completely wrong. One researcher tried passing infrared light through a tube containing the contemporary proportion of carbon dioxide. As Tyndall had found, a certain amount of the light was blocked. But then the researcher doubled the proportion of carbon dioxide, and nothing changed. The same amount of infrared light disappeared into the gas.

How could that be? Surely adding more carbon dioxide should mean more of the infrared would be blocked. However, carbon dioxide turned out to be unexpectedly choosy about which frequencies of light to soak up, going for only a few restricted 'colours' of infrared. So restricted, in fact, that the merest trace of carbon dioxide could absorb all the available light in these colour ranges. After that, you could double, triple, even quadruple the amount of carbon dioxide in the tube and the rest of the infrared light would still pass through untouched.

Soon, other objections began to emerge. The oceans contain huge quantities of carbon dioxide – fifty times as much as the atmosphere. Most of the extra gas put out by factories would surely be taken up by that vast reservoir, leaving behind only a minuscule amount to slip into the air.

All in all, these comforting notions sat well with the prevailing picture of a world in which nature was vastly stronger than the puny forces of mankind, and where natural cycles somehow balanced everything in the end. There was nothing to worry about, or even to be particularly interested in. Extra carbon dioxide couldn't possibly warm the planet. Or so it seemed.

In the decades that followed, a few researchers kept up their interest in the climatic effects of carbon dioxide. Some were just vaguely curious, others convinced there was something in Arrhenius's idea, and among them all they just about kept the issue alive. Meanwhile, the cities of the world began to spread. Lifestyles in many countries started to shift from the grinding slog of agrarian societies to the glories of the industrialised world. Year after year, more factory chimneys were springing up to pour carbon dioxide into the air. Then there were railways, motorcars and jet engines, and what had been a trickle of carbon dioxide became a flood. Between Arrhenius's time and the end of the twentieth century, Earth's population would more than quadruple, and the average use of energy by each of these people would also quadruple. The rate that carbon dioxide poured into the atmosphere from human activities would multiply a staggering sixteen times. Nobody guessed this would happen – how could they? For the time being, the growing amounts of carbon dioxide in the air went unnoticed, and unsung.

Then, in 1952, one of the central criticisms of Arrhenius's work unexpectedly crumbled. Adding carbon dioxide was supposed to make no difference because the amount already present in our air appeared to be grabbing all the infrared radiation that it could. But new measurements and theory began to suggest that this argument was seriously flawed. Those early experiments were made in an ordinary lab, at ordinary temperatures and pressures. But aloft, where most of the infrared fielding takes place, the air is frigid and thin. That, it turns out, changes everything. At such low pressures and temperatures, carbon dioxide no longer soaks up every bit of radiation in its favoured colours.

This new finding inspired a weapons researcher at the Lockheed Aircraft Corporation. Gilbert Plass was an expert on infrared radiation – he spent his days using it to try to develop heat-seeking missiles. But in the evenings, Plass enjoyed reading about science more generally. When he came across Arrhenius's discredited theory about carbon dioxide and infrared light, he became curious about how much of a difference the new results would make. Fortunately, he had no need to resort to months of calculations with pencil and paper; thanks to his day job, he had access to one of the newly invented digital computers. Working mainly in his spare time, Plass fed in the revised figures. The result was just as he had expected: adding more carbon dioxide to the air *can* make a difference after all, and the effect on climate looked to be significant.[18]

The next notion to fall was the idea that oceans would soak up most of the carbon dioxide. Researchers began to realise that the warm surface layer of the ocean doesn't mix very much with the colder water underneath, which means most of the carbon dioxide taken up by the ocean is quickly recycled into the atmosphere. Nobody could be exactly certain what difference this would make.[19] What they really needed was to know whether carbon dioxide levels in the atmosphere were actually changing. And if so, by how much?

That's when a young American researcher named Charles 'Dave' Keeling entered the scene. Keeling had read Plass's work and had discussed it with him. He was fascinated by carbon dioxide and what effect it might have on Earth's climate, and he became convinced that the only way to know for sure was to measure it. To do this, he developed instruments to measure carbon dioxide levels with delicate, extraordinary accuracy. Next, he placed them on the top of Mauna Loa, an extinct volcano on the Big Island of Hawaii, well away from the sorts of local industrial influences that could ruin his results. But he didn't want to measure for only a month, or even a year. He wanted to keep the measurements going indefinitely.

Keeling was inspired, technically brilliant and also – fortunately – bull-headed. Fortunately, because he discovered that there was no funding available for long-term studies like the one he had in mind. There was nothing wrong with making a few measurements once in a while, he was told repeatedly by the US science-funding agencies. But keeping highly expensive and very technical instruments ticking over constantly in Hawaii for *years*? There was simply no need.

Keeling, however, refused to hear the word 'no'. He wrangled and scraped and insisted, and he somehow managed to keep his instruments in place and switched on. It wasn't long before he was proved to be right. Even between one year and the next, he could see the difference in carbon dioxide levels. And it was exactly what you'd expect if the oceans were not, after all, soaking up those human outpourings.

Keeling continued making those measurements now for more than forty years. When plotted out on a graph, his 'Keeling curve' has become one of the most famous icons of the global warming debate. For as the years have passed, the carbon dioxide levels it shows look nothing like a flat line, or even a gentle rise. Instead, they rear up exponentially, like a malevolent tidal wave ready to crash.

But could carbon dioxide really be warming up the world? Sophisticated new computer models suggested that it should, but they struggled to come up with a consistent answer. Some said that doubling carbon dioxide levels would increase global temperatures by one degree, others by eight or nine. Perhaps the answer was to look at exactly how much temperatures really had risen, if at all. But here, there was another problem. Temperatures fluctuated perfectly naturally from year to year, and that made it very difficult to discern any possible warming from the thicket of ordinary highs and lows.

This is one reason that global warming researchers have always had an image problem. It's not too hard to jolt people into action if you can point to a massive oil spill, or a forest that's been devastated by acid rain. But where the effects of carbon dioxide are concerned, the long view is the only one that matters.

Nobody will *ever* be able to say 'this particular heat wave was caused by global warming' or finger it as the culprit for that individual flood. Instead, the potentially nefarious effects of carbon dioxide are all about something that's much harder to pin down: *trends.*

And yet the world was now stirring to this new threat. Records seemed to suggest that temperatures had risen by a fraction of a degree in the past century, and though it wasn't by much, it was the first real sign of change. Then, in 1995, an international group of climate scientists announced for the first time that the balance of evidence, in their opinion, had slipped over a threshold. Global warming, they declared, is upon us. Hot on the heels of that announcement came news that 1995 was the warmest year since records began. The year 1997 was even warmer, and 1998 warmer still.

And then, a scientific paper published in 1999 struck what many consider to be the killer blow against global-warming sceptics. The paper came from decades of work in what is, officially, the coldest place on Earth. Vostok station, a Russian base in the frigid heart of the Antarctic Ice Sheet, reaches winter temperatures cold enough to shatter steel. Even in summer it's a forbidding place. The temperature scarcely ever rises above –25 degrees Celsius and the air is almost as dry as the Sahara. Its handful of occupants live in a station that is perpetually starved of funds and seems to cling to the ice through sheer Russian tenacity.

But the ice at Vostok is miraculous. More than three kilometres thick, it holds a frozen archive of past climate stretching back hundreds of thousands of years. For decades Russian scientists, aided by some French and then American researchers, had been drilling a hole into this storehouse, and the deeper they went, the farther back in time they penetrated. They had already announced a record of temperatures for the past 400,000 years and discovered a series of four successive ice ages, each with a warmer period in between. But what they produced in 1999 caused a sensation. They had managed to recover not only temperatures, but minuscule amounts of Earth's ancient atmosphere.

How could something as insubstantial as air be preserved? Well, whenever snow falls at Vostok, it traps a small amount of air among its flakes. Gradually, over the years, the flakes become buried by yet more snow. They are squeezed and compressed until finally they turn into ice. At this point, the trapped air can no longer wriggle its way to the surface. It remains in cold storage, tiny bubbles that provide time capsules of the planet's ancient atmosphere. The researchers at Vostok had not only managed to recover these tiny bubbles. They had carefully broken into them and released air last breathed when the human species, *Homo sapiens*, had only just appeared on the evolutionary scene.

And then they had measured it. With extreme patience, the scientists managed to extract these tiny quantities of carbon dioxide and feed them through their measuring devices. They produced a record of carbon dioxide levels stretching back 400,000 years, to match the one they had already created for temperatures.

Plotted side by side, these two records revealed something remarkable. Whenever the temperature was lower, so were the carbon dioxide levels. Whenever the temperature was higher, the carbon dioxide was higher, too. Climate and carbon dioxide clearly marched in lockstep. Tyndall and Arrhenius had been absolutely right.[20] We still don't know the exact connection between carbon dioxide and temperature, or all the complex interrelations of Earth's atmosphere. But history shows us that carbon dioxide is clearly a hugely important driver for our planet's temperature.

And there was something else, something even more striking. Carbon dioxide levels seemed to vary quite naturally, along with natural changes in temperature. But when the researchers studied their record more carefully, they discovered that at no point in the last 400,000 years had carbon dioxide levels been anything near what they are today.

A newer ice core, drilled a few hundred kilometres from Vostok at Dome C by a consortium of European researchers known as EPICA, has now gone even farther back in time, almost 800,000 years. They found exactly the same story. Carbon dioxide changes mirrored temperatures with astonishing fidelity. And as far as they could reach with their ingenious frozen time-machine, levels in our atmosphere have never been as high as they are today. The highest level Earth managed naturally during that time, which includes all of human history, was about 280 parts per million, or 0.0028 per cent. But today we have more than 380 parts per million – and it is rising.

Nobody yet knows what effect this will have on our world, although most scientists think that it's now too late to avert at least some amount of change. We know, or at least suspect, that in its ancient history our planet experienced levels of carbon dioxide even higher than today's. But that was long before humans, or even our ape-like ancestors, existed. In the past few hundred years, we've put a huge amount of effort into developing our society according to the present climate, the present pattern of floods and storms and rainfall, of crops and livestock. We are embedded in our present homes and places of work. And we can't just lift up our skirts and move if the warming sea begins to rise and encroach on our waterside cities, if storm surges begin to devastate our coastlines, and if the interiors of our continents begin to turn into dust bowls.

Meanwhile, yet more evidence has emerged from the ice, suggesting that our entire complex climate system, driven by the engine of Earth's atmosphere, can sometimes be delicately balanced between dramatically different states. One slight shift can send temperatures soaring or plummeting.[21] In 1987, an ever-prescient climate researcher from New York, Wally Broecker, commented that we had been treating the greenhouse effect as a 'cocktail-hour curiosity', and it was time to take it seriously.

The climate system, he said, was a capricious beast, and we were poking it with a sharp stick.

After 35,000 people died during a fierce heatwave in Europe in 2003, the UK government's chief science adviser declared that global warming was 'an even worse threat than terrorism'. But while politicians wrangle and scientists plead, we continue our lives more or less as normal. And every time one of us drives a car, catches a plane, switches on an electric light or does any one of a myriad of ordinary tasks, another whiff of carbon dioxide rises up into the sky.

One final cautionary tale about the powers of carbon dioxide comes from our sister planet, Venus. Being a little closer to the sun than we are, you'd expect Venus to be slightly warmer, but in many other ways – size, for instance – it could be our twin. However, at some point in the past carbon dioxide worked its wicked magic on Venus's air. For some reason, a little too much carbon dioxide trickled out from Venus's volcanoes into its atmosphere. The air grew warmer, which meant it sucked up water from the oceans. The extra water vapour acted as a greenhouse gas in its own right and reinforced the behaviour of the carbon dioxide. Soon the atmosphere was filled with carbon dioxide and water molecules, all catching infrared heat as it tried to escape and flinging it back to the ground. The result: Venus's oceans are long gone. The rocks on its surface are now dry as a bone, and hot enough to melt lead.

Many researchers take comfort from Venus's greater proximity to the sun and say that such a greenhouse catastrophe could never happen here on Earth. But there is a chance they might be wrong. A recent project[22] that used thousands of PC screensavers to run versions of a climate model and predict the possible future outcome of climate change suggested a doubling of carbon dioxide levels could produce a global temperature change as high as 11 degrees Celsius. That would trigger such droughts and wildfires that yet more carbon dioxide would flood into the atmosphere, leading to a catastrophic meltdown.

The chance may be small, of the order of 1 per cent, but it is still possible.

Carbon dioxide, then, is both a crucial and dangerous element of the air. We need it for food and warmth, but we abuse it only at our peril. Along with oxygen, nitrogen and the sheer thickness of the air, it helps transform our lump of rock into a living, breathing world. The final part of this transformation involves not what the air contains, but how it moves. You encounter the motion of air every time you're struck by a gust of wind, and every time a door mysteriously slams in the house when windows are open on either side. But it is in the grand movements of vast bodies of air around the planet that wind truly comes into its own as an agency for life.

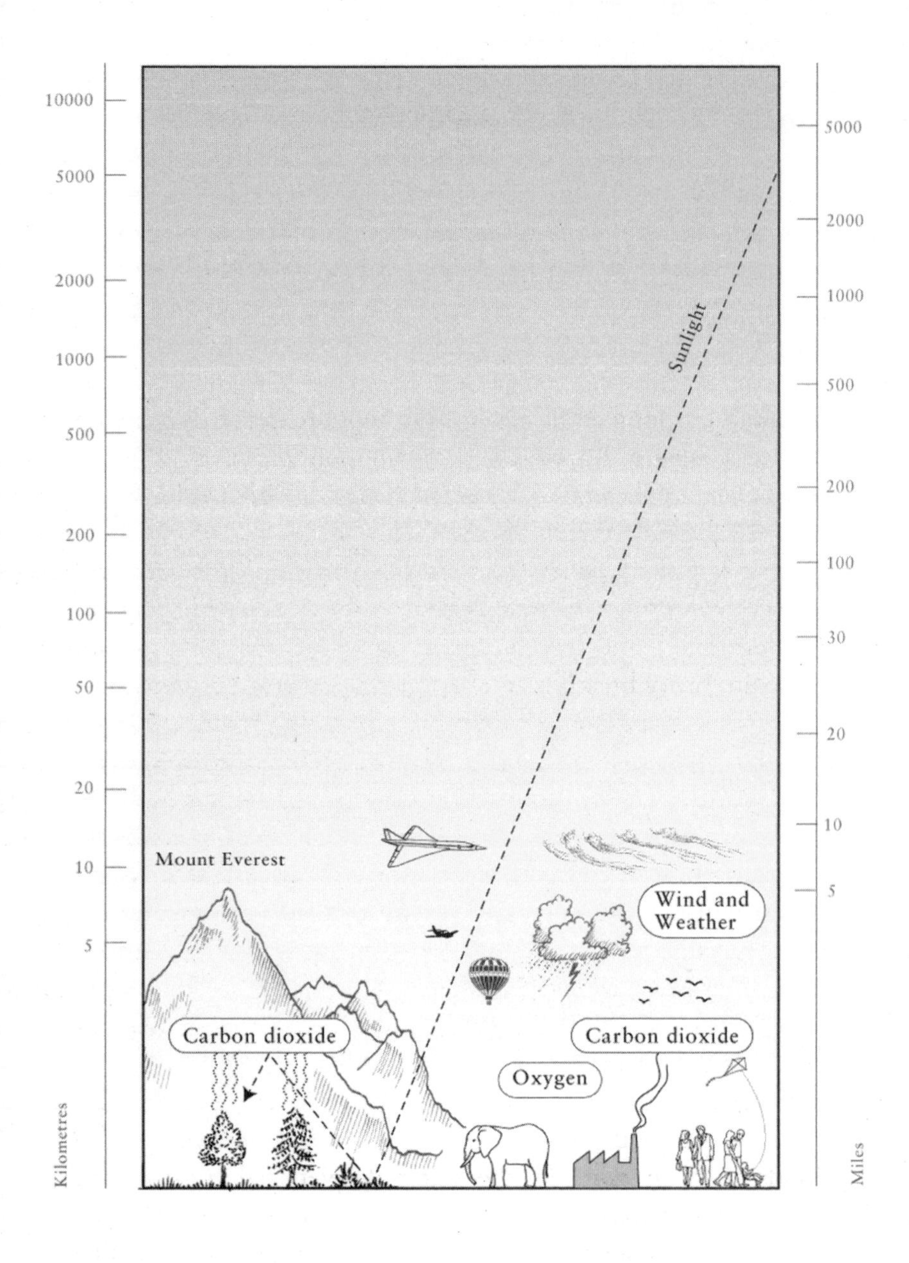

10000
5000
2000
1000
500
200
100
50
20
10
5
Kilometres
5000
2000
1000
500
200
100
30
20
10
5
Miles
Sunlight
Mount Everest
Wind and Weather
Carbon dioxide
Carbon dioxide
Oxygen

# *Chapter 4*

# BLOWING IN THE WIND

For almost as long as air has been in motion, living things have hitched a ride on the wind. The obvious animals to find in our ocean of air are the ones that can fly. But aside from the birds and the bees, there are plenty of other living things that simply float. The air is full of pollen grains seeking to fertilise plants and protect them from inadvertent incest; there are seeds hunting out new fertile soil, and tiny-shelled sea creatures whipped up with the foam. Every breath you take contains dozens of microscopic fungi, not to mention the minuscule viruses and bacteria that are spreading their secret infections. Even before anyone knew about microbes, there were those who suspected that air could bring disease – hence malaria ('bad air'). Every word you speak, especially those with explosive consonants like *p* and *t*, sprays bacteria out into the air around you, ready to be passed on by the wind. A cough produces 2,000 and a sneeze, 400,000.[1] Certain devious viruses have even evolved so that when we have bred them in our bodies, we spray them out with a sneeze so they can fan out on the wind.

Other bacteria hitch a ride on clouds, and may even choose their own drop-off point by making ice crystals that induce the clouds to form rain. As the water droplets fall back to the ground, the bacteria simply fall with them.[2] Orb and crab spiders spin invisible gossamer threads and then use them like sails to catch the wind. The slightest updraft, the thinnest shaft of sunlight to warm a patch of air, and the spiders will launch themselves into the air to begin their travels. Nobody yet knows exactly how they arrange their journeys. They may simply continually land and relaunch themselves until they find the ideal resting spot, though some scientists think they might regulate their flight, reeling in threads to raise or lower their sail, or perhaps even steer.

And of course, the wind carries people. Even before the days of balloons and planes, wind was the only way to cross the seas. In fourteenth-century Europe, after centuries of dark ages, blighted by the battles of the Middle Eastern crusades, the Renaissance had dawned, and with it a new urge to look outwards. This was the age of the great ocean explorers, and their fortunes rested on the wind. Some seventy years before the birth of Galileo Galilei, another Italian, a former weaver from Genoa, knew how important currents of air would be for his mission. What he didn't know is that he was about to stumble across two of the biggest wind systems in the world, the trade winds and the mighty westerlies, great globe-girdling torrents that form part of the final crucial ingredient for life on Earth.

**3 August 1492**

Half an hour before sunrise, a small fleet slipped out of the Spanish port of Palos. Two of the ships, the *Pinta* and *Niña*, were caravels, with small triangular sails. But the flagship, the *Santa María*, was magnificent, square-rigged with castles fore and aft. She was gaily painted above the waterline, her sails plastered with crosses and heraldic devices. The royal flag of Spain hung from her main mast, while the foremast bore the expedition's own banner, a green cross set on a white background, bearing four gold crowns.

The commander of the *Santa María* had already had several names in his forty-one years, and would be given more in the centuries to follow. His Genoese parents had known him as Cristoforo Columbo, but he had shaken off his Italian roots and language along with his father's profession of wool-combing. Now he was a seaman, a captain general on a mission for Spain. He had embraced his adopted country with typical fervour. He wrote in Spanish, even in his most private journal, and the name he used was a Spanish one: Cristobál Colón.

The man we now know as Christopher Columbus looked nothing like his Iberian crew. His hair had once been tawny, but it had turned snow white ten years ago, when he was barely thirty. His face was pale and freckled, his nose Roman, and his blue-grey eyes often burned with passion, and with temper.

His mission, of course, was to sail west in order to reach the East. Fifteenth-century Europe was full of tales of the fabulous riches of the Orient. In the previous century the Venetian voyager Marco Polo had written a graphic (though embellished) account of his travels in these lands of spices, silks, gems and unimaginable quantities of gold. The newly invented printing press had spread his stories throughout Europe; merchants and monarchs read Marco Polo's book and felt their fingers itch. There had to be a way to reach those cargoes.

But the countries that Marco Polo had called Cathay and Cipangu, and that we now know as China and Japan, remained stubbornly out of reach. The land journey was far too long and perilous for transporting costly merchandise, and the entire continent of Africa was in the way of the eastern sailing route. Thus the whisper began. What about travelling westward? If you could slip across the ocean and approach the Orient from its backside, all the wealth and glory waiting in those distant lands would be yours.[3]

Now at last, after years of fundraising and pestering, Columbus had his chance. His backers, King Fernando and Queen Isabella of Spain, had provided him with this handsome fleet and promised him the rank of Admiral of the Seas if he should be successful.[4] There was only one thing more he needed: a wind to blow him westwards.

Apart from the islands of the Azores, discovered a few decades earlier by Portuguese sailing ships, the Iberian peninsula marked the western edge of the world. Beyond was the stuff of legend: some spoke of a fabled island called Antilla, supposedly discovered by the Carthaginians; others of fragments of Atlantis, which had somehow escaped inundation, or of a gigantic and beautiful island bearing seven great cities, each more splendid than the last. Many had tried to explore these lands, but until now all had been beaten back by a fierce headwind that whipped up the seas to a fury. The wind blew from exactly the wrong direction – the west – and no sailing ship could pass.

But Columbus had a plan. In years past, when he was earning his sailing spurs, he had made several voyages down the African coast. And whenever he passed the Canary Islands, especially in winter, his ship had been buffeted gently but steadily from the east.

This was the wind that Columbus had resolved to try to catch, the one that he hoped would carry him at least some of the way over the western ocean. When his three ships left Palos they headed not west, but south.

The journey to the Canaries was difficult, the wind fitful and contrary. Life on board settled down into its routines. Crew and captain both were deeply religious. Each turning of the hourglass was accompanied by a boy singing out a blessing. The days began with hymns to Mary and finished with evening prayers. Only the captain had a tiny wooden cabin. Since hammocks had not yet been discovered (they were waiting in the Caribbean), the rest of the crew had to find berthing where they could on deck, tying themselves in against a sudden rolling of the ship.

Columbus was fretful. He wondered constantly if the easterly winds that he sought would even exist, and, if so, how far west they would take him. (By exasperating convention, meteorologists label a wind by where it's coming from, not where it's going to. So an 'easterly' blows from the east towards the west, and was just what Columbus needed.) The small fleet reached the Canaries three weeks after the start of their voyage. They re-provisioned, and then, on 6 September, the ships weighed anchor and turned full west.

For the whole of the next day, the ocean held its breath. And then, on Saturday 8 September, up rose a wind from the east and Columbus found himself where he had always wanted to be: in uncharted waters.

The new easterly wind was beyond anything Columbus had dared hope for. Over the next two weeks, it blew the fleet steadily farther and farther westward towards their goal. The sailing was magnificent. The weather, Columbus noted in his journal, was like Andalusia in spring. 'The mornings are most delightful, wanting nothing but the melody of nightingales,' he wrote, and a few days later: 'The sea is smooth as a river, and [there is] the finest air in the world.'[5] The speed of the small fleet was astounding. On their best day's run they covered 182 miles, at an average speed of eight full knots. And the wind never stopped coming.

Columbus had no idea what he had found, but this ultra-reliable easterly would prove to be every bit as important a discovery as the New World to which it carried him. For it was one of two giant belts of winds that stretch around the globe in the tropics, one north and one south of the equator. They are so steady and diligent that they would come to be known as the 'trades', for the safe trading routes that they made possible. (They may even have been used by humans before Columbus. Norwegian archaeologist Thor Heyerdahl has shown that the trade winds could blow a simple sailing vessel made of reeds from Europe all the way to the Caribbean, and suggested that the idea of building pyramids could have reached Central America by ancient Egyptians who took this route. Though if the ancient Egyptians did reach the Americas with tales of their pyramidal technologies, you'd think that they would also have mentioned the wheel.)

For Columbus, however, the trades were beginning to become too much of a good thing. They were so steady, and so unabating, that his crew began to grow nervous. He'd had enough difficulty finding sailors willing to make their way into the unknown, to equip their ships with a year's provisions at a time when even the most daring voyage lasted only a few weeks. Now, as the fleet sped its way west, murmurs of unease started to spread. This wind that was blowing them with such speed and efficiency never seemed to die down. Columbus noted grimly in his diary that his crew had 'grown much alarmed, dreading that they never should meet in these seas with a fair wind to return to Spain'.

Columbus did his best to distract his crew from their fears. He pointed out any scrap of evidence, however dubious, that made them seem close to land, and recorded each in his journal. The 'signs' could be almost anything: 'It drizzled without wind', or 'a great mass of dark, heavy clouds appeared in the north', or 'saw a whale, an indication of land, as they always keep near the coast'. He even announced false distances when announcing the day's progress, on the dubious principle that this might help. ('Sailed this day nineteen leagues, and determined to count less than the true number, that the crew might not be dismayed,' he wrote on Sunday, 9 September, and again on the tenth: 'This day and night sailed sixty leagues . . . reckoned only forty-eight leagues, that the men might not be terrified.')

What he really needed was land. And this he found, famously, in the early morning of Friday, 12 October. Seaman Rodrigo de Triana, lookout on the *Pinta*, was the first to spot the cliffs of San Salvador, crying out '*Tierra! Tierra!*' Like every other crewman, he was hoping for the reward of ten thousand maravedis – a decent annual salary for an able seaman – which the queen had promised in perpetuity to the first to sight new lands. Columbus, however, maintained that he had seen a light 'like a wax candle moving up and down' some hours earlier, and claimed the reward for himself.

With the morning light, Columbus and his companions became the first Europeans to step on to a new world. Although nothing he saw on land bore any resemblance to Marco Polo's descriptions, Columbus always believed that he had found the Indies. But he had to revise his ideas of sophisticated treasures, focusing on the aspects of the world that were riper for exploitation: the cotton, the wood, the spices and the surprisingly gentle people. These seemed almost as if they came from Eden before the fall. They were naked, open, curious and – so Columbus noted in his diary – had so little idea of weapons that, when he showed them a sword, they picked it up by the blade and cut their hands on it. 'It appears to me that the people are ingenious, and would be good servants,' he wrote, 'and I am of the opinion that they would very readily become Christians.'[6]

Still in search of riches to justify his voyage, on 23 October Columbus decided to head for an island the natives called Cuba, where he hoped at least to find 'much profit . . . in spices'. Cuba did indeed yield plants that would provide people in the future with great profit. The natives there had the peculiar habit of rolling up herbs inside dried leaves, setting them alight and inhaling the smoke, which turned out to be more pleasurable than it looked. Las Casas, Columbus's friend and the transcriber of his journal, wrote his own account of this practice: '[The herbs] are dry, and fixed in a leaf also dry, after the manner of those paper tubes which the boys in Spain use at Whitsuntide: having lighted one end they draw the smoke by sucking at the other, this causes a drowsiness and sort of intoxication, and according to their accounts relieves them from the sensation of fatigue.' Prefiguring later attitudes to this newfound weed, Las Casas himself was censorious, adding sternly, 'I do not see what relish or benefit they could find in it.'

By early January, Columbus decided he had enough gold artifacts and specimens of exotic spices and woods – not to mention natives – to impress his royal patrons, and he decided he should head for home. His ship, the *Santa María*, had accidentally run aground, so Columbus decided to leave her and a handful of men, to begin a colony. He claimed the *Niña* for himself, and on Wednesday, 16 January, the two caravels began for home.

Immediately they faced the problem that had been so feared by Columbus's crew on the outward journey. The winds that had taken them so steadily to the 'Indies' were now blowing in their faces. Against these headwinds, how would they ever get home?

The *Pinta* and *Niña* were forced to beat against the prevailing trades, creeping ever northwards so they could edge their way east. Farther and farther north they crept, until, out of nowhere, came a miracle. On 31 January the wind swung. Suddenly a gale filled the sails of the two caravels and pointed their prows towards Europe. The ships found themselves running before a wind that seemed to be urging them homewards, hour after hour, day after day, their sails taut, racing over the ocean at giddy speeds: nine, ten, even eleven knots.

Once again, Columbus had made a discovery to rival that of the Americas. For this new wind was another part of the global conveyer belt, and the natural complement to the easterly wind that had borne him there. He had stumbled across the mighty Earth-encircling westerlies. Like the trades, they, too, appear in each hemisphere. The southern ones are responsible for the famous 'roaring forties' around 40 degrees latitude, and the infamous storms that have long plagued sailors around Cape Horn.

The northern ones have also claimed plenty of victims, for the westerlies are nothing like the gentle, steady trade winds. They are fierce and furious. At first, Columbus's ships gamely weathered their buffeting, their crews thrilled to be moving homewards so rapidly. But on 14 February the winds let rip. They worked themselves up to a frenzy, whipping up the water and slamming it into the leaking wooden hulls of the two small ships. 'The sea was terrible,' Columbus wrote, 'the waves crossing and dashing against one another, so that the vessel was overwhelmed.'

The crew did the only thing left to them: they prayed. And among their prayers they made many vows, private and public, as to what they would do if saved. Some of the promises were extraordinarily specific. They cast lots using a hat filled with dried peas, one for each crew member, to decide who should swear to make a pilgrimage to St Mary of Guadalupe carrying a wax candle five pounds in weight. Columbus himself drew the pea clumsily marked with a cross, and immediately made his vow. There were more lots, more pilgrimages promised, and every crew member swore to go in procession, 'clothed in penitential garments', to the first church dedicated to Our Lady that they should encounter.

Columbus's preparations were practical as well as metaphysical. Fearing that, should they perish, all record of their journey would be lost, he braced himself against the lurching ship long enough to write a secret account of his adventures, to be delivered, if found, to the king of Spain. This he rolled in a wax cloth and placed inside a wooden cask before hurling it into the sea, an act that his crew took to be another, albeit bizarre, sign of devotion.

In the end, of course, the westerlies had mercy. After a few more hours of horrors, the storm finally abated and Columbus limped back to Spain. The tales he took with him were to change both Europe and America immeasurably, though not many of the people who had made this first contact would benefit from it. Separated from the *Niña* by this great February storm, the treacherous captain of the *Pinta* tried to race Columbus to the king and queen, to be the first with the news. But he arrived just too late. He took to his bed immediately, crushed with disappointment, and died within the month. Columbus fared a little better – his is the name that survives in the history books and memorial days. But even he didn't long hold on to the titles and wealth with which the monarchs showered him; his unsuccessful attempts to govern the people he had discovered would turn the Spanish crown against him, and his final return from the 'Indies' two voyages later would be in chains. The gentle natives he had encountered, meanwhile, would gradually learn the true horrors to be found at the hands of these men whom they thought had come from Heaven.

But while Columbus's New World changed through its contact with the old, the gentle trades and boisterous westerlies that had carried him blew steadily on. They still do so today, and, as long as Earth has air to feed them, they always will. And as they do so, they transform our world.

Nobody in Columbus's time had the slightest notion how far-reaching the winds he had stumbled across would prove to be. It would be some time before mariners even realised that they encircled the globe, and longer still before the first tentative suggestions emerged for why they should exist. But the full explanation of their powers would have to wait four hundred years, for a shy farm-boy genius scratching out a living in the dirt of the continent that Columbus once claimed for Spain.

**Spring 1831**
**Berkeley County, West Virginia**

In many ways the farm was a good thing. William Ferrel's father had bought it two years ago, and life there was more settled than in the erratic lumber trade. Also, there was plenty of space for the young Ferrel to slip away and think. In his boisterous family of six brothers and two sisters he could easily remain unnoticed, the quiet one in the corner, lost in his own thoughts.

The problem was that there was nothing for him to read. Ferrel was fourteen. For the past two years he had picked up what he could of reading, writing, counting and grammar, huddled together with the other farm children in the freezing school hut. Perhaps it would have been pleasant in the summer, but then the daylight hours were too long and precious to be wasted on learning and even the youngest children were needed in the fields. Studying was for the winter, when an icy wind slipped beneath the white oil paper that was tacked over the windows in place of glass, and crept through the gaps in the cabin's rough-hewn logs.

The cold hadn't bothered Ferrel much. What he minded more was that school was now over for him. It was time to get on with the farm. And yet his mind wouldn't stop working. He was desperate for something, anything, to read. The family received a tiny local newspaper, the *Virginia Republican*, which was published every week in the nearby town of Martinsburg. Ferrel pounced on this the moment it arrived, scouring it in search of some rare article that might let his mind work.

Then he saw a copy of a book that he immediately coveted. It was called *Parks Arithmetic*, and it contained enticing diagrams for how to calculate the circumferences of shapes and their sizes, too. He wanted it desperately.

Still, Ferrel was too shy to ask his father for the money – for a book, of all things. Instead, he waited until he managed to earn fifty cents by helping at a neighbour's farm during harvest, and then headed off to the bookshop in Martinsburg. The book, it turned out, cost sixty-two cents, but the kindly storekeeper let him have it anyway.

*Parks Arithmetic* was the unlikely beginning of Ferrel's lifelong love affair with books. He devoured the text, working eagerly through its problems and delighting at each answer that he found. Arithmetic came easily to Ferrel. It was theoretical, even perhaps imaginary, and held no apparent connection to the natural world that he lived and breathed on the farm. But he loved it the way crossword addicts love their puzzles. Give him the problem, and he would find the solution.

Then, on the morning of 29 July 1832, something important happened to connect this solving of puzzles to the world around him. Ferrel was on his way to the fields when he saw an eclipse of the sun. He hadn't expected it, but he realised that somebody must have known it was coming. The moon was perpetually floating over his head, and once in a while it must get between Earth and the sun and briefly block the view. A lunar eclipse had to be the same sort of thing, except that the moon's view of the sun was being blotted out by our shadow. In each case, the cosmic do-si-do of the planets had to be predictable.

Of course, Ferrel had never studied astronomy. He didn't know the shape of the moon's orbit, and in any case hadn't learned enough geometry to be able to calculate its path. But he could look for patterns. If he worked hard enough, with the only tools he could find – an elementary geography book containing information about the globe, and a farmers' almanac predicting the positions of the sun and moon at different times of the year – perhaps he could work out the times and dates of future eclipses.

This was a fabulous new puzzle, one that appealed to his practical streak as well as to the theoretician in him. He worked every moment that he could spare from his chores, day and night, carefully inscribing his efforts in a notebook. (At one point, he almost gave up in despair. He had wrongly assumed that Earth's shadow would always be the same diameter as Earth itself, whereas in fact it gets steadily smaller as you move farther away. With this error in place, his geometry simply wouldn't make sense. Then, one evening on the threshing floor, he noticed that a shadow cast by a wooden plank was thinner than the plank itself, and he raced back to redo his calculations.)

After two years of hard labour, Ferrel finally had his predictions: the following year, 1835, would have one solar eclipse and two lunar ones. He had no need to wait for the specified dates and times to find out if he was right – the calendar for 1835 would have the answers. And when it arrived, Ferrel was triumphant. The three eclipses were due exactly as he'd predicted, and his times were accurate to within a few minutes.

Now Ferrel was hooked. A neighbouring youth told him about a book he had seen that contained 'a great many diagrams' and was on a subject called trigonometry. Back at the Martinsburg bookshop, Ferrel bought the nearest thing he could find – a surveying text – and began studying it avidly.

He had hardly any time spare that summer – he was supposed to be working all the hours of daylight on the threshing floor, separating wheat from its chaff. Luckily the building had large doors at either end, made of wide planks of soft poplar wood. With this at hand, Ferrel had no need of a blackboard or of paper and pen. He drew his diagrams on the doors, making circles with the prongs of a pitchfork and straight lines with a single prong, using a small piece of board as a ruler. (The lines he had carved survived wind and weather for several decades, and even after he had become an exalted scientist, each time he returned to visit the farm he went to look at them.)

That winter, Ferrel borrowed another geometry text from an old surveyor who lived in the mountains and studied it by the feeble light of a tallow candle, or more often by firelight. He had a stock of light wood, and each log he threw in would encourage the fire to flare up, though only for a few minutes at a time. The next winter, he rode for two days through the snow to buy a copy of *Playfair's Geometry* from Hagerstown in Maryland. The more he learned, the hungrier he became.

Ferrel wasn't simply studying to know what others had already figured out. He now felt an urge to discover things, to explain Earth in ways that hadn't been explained before. Thanks to his work on the eclipses, his favourite puzzles had become the ones that truly existed, in the real world that he sensed around him.

With money made from teaching, and some donated by his bemused but supportive father, Ferrel attended college, where he studied algebra, geometry and trigonometry. (Finding this didn't fully occupy his intellectual energy, he also picked up Latin and Greek grammar on the side.) In 1844, after a hiatus to earn more money for fees, Ferrel finally graduated, aged twenty-seven. He had gone from farmer to mathematician, but there were still few academic options for a poor boy from West Virginia. He returned to his day job of teaching and devoted his evenings and all his spare time to research. Always, he was seeking the next subject that would fire his imagination with the fervour he could scarcely control.

A decade passed, Ferrel working at this and that in between his teaching. And then, in 1855, at the age of thirty-eight, he came across a book called *The Physical Geography of the Sea*, by Lieutenant Matthew Fontaine Maury of the US Navy. The book was a curious one. It contained tables and tables of data on winds, currents and air pressures collected from around the world. But it was also full of what seemed to be odd theories about how these numbers connected. Ferrel bought the book and took it home for closer study.

Ferrel didn't know it, but Maury was already famous, or more accurately infamous, in the nation's capital. He was an ambitious, bombastic military man with apparently limitless energy to promote himself. He had made his name through the undoubtedly excellent idea of collecting logbooks from ocean-going vessels, tracing their routes and collating their wind records so that he could publish maps of the prevailing winds. The resulting *Charts of Winds and Currents* had been an immediate hit. Unfortunately, it had also led Maury's already considerable ego into believing that he was a great scientist. He was convinced he was now qualified to speak with scientific authority on every imaginable subject. And when, in spite of having no background whatsoever in astronomy, he managed to get himself appointed superintendent of the US Naval Observatory in 1844, he became insufferable.

Although he wasn't an easy man to like, you could feel sorry for Maury. He wanted nothing more than to be one of the scientific gang. But his problem was that he simply wasn't very good at science. His theories were wild. He invoked random magnetic forces to explain phenomena he couldn't begin to understand, and when that failed, he resorted to thundering passages from the Old Testament to justify his 'scientific' claims.[7]

Others in Ferrel's day were either scornful of or downright alarmed by Maury, especially when he began claiming to be an expert in meteorology and urging Congress to accept him as head of a new, and highly dubious, system for predicting America's weather. By 1856, the burgeoning scientific community had already begun openly referring to him as a 'humbug'. Maury was just as insulting in his ripostes. When he was criticised at a scientific meeting at the Smithsonian Institution in Washington, he responded by declaring that James Smithson, the institution's otherwise illustrious founder, had been born a bastard (a fact that everybody knew but nobody ever mentioned).[8] The city's principal newspaper, the *Washington Star*, then took up the cudgels, describing Maury's work as 'one of the most remarkable and successful careers of unblushing charlatanism known in the world's history'.[9]

Ferrel was sublimely unaware of all this Washington name-calling and wouldn't have paid attention to it anyway. But he was intrigued by what he read in Maury's book, *The Physical Geography of the Sea*. In this book, Maury had set out much of the data that he had collected from records of wind currents and pressures. But, in a bid to seem more scientific, he had also filled the book with his bizarre theories of how the winds work. What Ferrel read set the circles turning in his head. Somehow there had to be a way to make the connection between all the disparate winds that Maury was describing, one that Maury himself clearly hadn't found. It seemed a shame to waste this valuable data on such feeble ideas. What's more, Ferrel was sure the answer would involve his favourite subject: geometry.

He decided to take the book to one of his best friends in Nashville, a medic from the college named Dr William Bowling. Ferrel had no family in the city, and not many friends. He was far too shy to socialise with strangers, but the few people who had managed to break down his barriers had become very close to him. Bowling was one of these, and he especially loved talking to Ferrel about science. He was the publisher of the *Nashville Journal of Medicine and Surgery*, and had been trying for years to give this journal some intellectual clout of the sort that Ferrel always seemed to carry with him. Ferrel explained his interest in Maury's data and his disquiet about the conclusions in the book. When Bowling heard this he was gleeful. 'Write me a review for the journal,' he demanded. 'Pitch into him.'

But gentle William Ferrel had no intention of pitching into anyone. He decided instead to use Maury's data to come up with his own ideas of how the winds work.

This had two odd consequences. By accidentally turning Ferrel's gaze on to meteorology, Maury's contribution to the subject would indeed turn out to be seminal – though not in the way he had hoped. And one of the most important papers ever written in the history of the subject was about to be published in an obscure Nashville medical journal.

Ferrel decided to ignore Maury's theories completely and look for the facts that he had gleaned from shipboard measurements and reports. It seemed that the winds in the two hemispheres moved in mirror image. On each side of the equator were the steady trade winds, which blew consistently from the east. Beyond these was another pair of wind belts, which were much stormier than the trades and usually blew from the west. Between these two sets of winds lay a mighty mountain range made not of rock, but of air. According to Maury's data, for some reason, air piled up in two giant ridges of high pressure, which circled the entire planet north and south of the equator, and separated the trades from the westerlies. This was the puzzle that Ferrel determined to solve: what drove these giant belts of wind to form, and why did the air pile up in between them?

Ferrel began by considering how moving air would be affected by passing over a surface (the planet beneath) that was itself moving. He picked up his pencil and began his calculations.

The maths was fiendishly complicated, but the answer surprised Ferrel with its sheer simplicity. In response to the turning Earth below, air itself felt the urge to turn, and always in a certain direction. Put plainly, 'in whatever direction a body moves on the Earth's surface there is a force arising from the Earth's rotation which deflects it towards the right in the northern hemisphere, but the contrary in the southern'.[10]

In other words, Ferrel had discovered the precise nature of the giddying effect that our planet's spin has on the air above it. Unscrupulous locals crouching over buckets at the equator might have tried to con money from you by convincing you that water goes down plugholes anticlockwise in the north and clockwise in the south because of it. This isn't true – in something as small as a bucket, any tiny perturbation in the water will be big enough to swamp the effect.[11] But on a larger scale, the effect of the planet's spin certainly does make storms circle in opposite directions in the different hemispheres. Chances are you think of this as the 'Coriolis effect'.

Nobody really understands why the effect was called this. Around 1930, forty years after Ferrel's death, textbooks began inexplicably to refer to it this way, after the French mathematician Gustave Gaspard Coriolis, who in 1836 had published a set of equations explaining how objects behaved in theoretical rotating systems. His maths was flawless. But Coriolis had never applied his research to the atmosphere, nor even imagined using it to explain the winds.

Even in Ferrel's time, meteorologists called the 'north hemisphere turns right, southern hemisphere turns left' rule after someone else. A year after Ferrel came up with his rule, Dutch scientist Christophe Buys Ballot from the Royal Netherlands Meteorological Institute published a paper simply pointing out the observation that northern air tends to move right. He had made no attempt to derive it mathematically, nor did he take it any further. However, because nobody had heard of William Ferrel, researchers began calling this 'Buys Ballot's Law', and the name stuck.

As soon as Buys Ballot heard of Ferrel's work, he was thoroughly embarrassed. Obviously Ferrel, not he, should have taken the credit. He even wrote to Ferrel offering that their names be used jointly. Poor, shy Ferrel made no attempt to hide his horror at this proposal. He wrote back immediately, concluding his letter with these words: 'Although I would esteem it a great honor to have my name in any way connected with yours, yet I will never encourage the change which you so generously propose.'

But to my mind (and to that of many others), the name of this effect belongs much more properly to this farm-boy genius. Though he was self-taught, and is still – thanks to his own diffidence – unsung, Ferrel both discovered the Coriolis effect in its fullest form and then went on to apply it. Using it, he was to become the first person in the world who would truly comprehend the winds.[12]

To understand what I will now obstinately – and unilaterally – call the 'Ferrel effect', think of the shape of Earth: a sphere spinning around an axis that goes through its centre. Although every part of the planet rotates precisely once a day, some parts have farther to travel than others. The equator has the hardest task. Being the broadest part of Earth, it has much the farthest distance to cover in its twenty-four hours, and every point on its surface is perpetually hurtling through space at more than 1,500 kilometres per hour. Farther north or south, the planet is narrower, and the speed of travel much slower; by the time you reach the poles, the surface doesn't move at all.

Air is affected by this because it is in contact with the spinning ground, and yet is free to move relative to it. The Ferrel (Coriolis) effect isn't a force so much as an optical illusion, brought about because we forget that we, too, are spinning with the ground beneath our feet.

Usually, this effect is explained in terms of motion north or south. One way is to take an orange to represent the spinning Earth and a black pen to show the motion of air travelling south. Start with the pen at the orange's 'north pole' and then move it directly south, while making the orange spin west to east. You will find that the black line you draw curls off to the west, to the pen's 'right'.

Perhaps a better way to understand the effect, rather than simply to see it in action, is to think about what happens if a piece of tropical air (with fairly slow spin, because Earth there is fairly narrow) starts to move south towards the equator (where the local spin is much faster because the planet is broader). When the air arrives at the equator, it finds itself in the fast lane, where the ground is zipping eastward under its feet. Here at the equator, our tropical air parcel lags behind so much it seems to be going backwards. In other words, it seems to turn to the west. To anyone standing on the surface, oblivious to Earth's spin, this would look exactly like air that was turning right.

The same principle applies to an air parcel moving north. Starting at the equator, it is already spinning very rapidly eastwards. But as it moves north, it finds itself hovering over a surface that is much more sluggish. Our air packet now finds itself in the slow lane, with its own foot still on the accelerator. And so it seems to turn eastwards, which is once again to its right.

This much had already been hinted at by a British scientist named George Hadley, who had been trying to explain the trade winds. But because this explanation rests on the difference in east–west motion between air and ground, until Ferrel came along everybody believed that it could only apply to motion north or south. A parcel of air moving eastward or westward relative to the ground should feel no effect at all.

And this was Ferrel's genius: through a combination of mathematical reasoning and brilliant intuition, he discovered that air is forced to curve even if it is moving east or west. In other words, whatever direction air is moving relative to the surface beneath, whether north, south, east or west, Earth's spin will always make the air appear to turn.

The explanation for why air moving east or west still turns is more complicated than for north or south, and much harder to picture. Think of how fast a particular piece of air is spinning west-to-east compared with the surface. Since Earth itself spins from west to east, any air moving towards the east will have all the spin of the planet beneath plus a bit extra. The faster anything spins, the more it wants to fly outward. (You can see this same effect if you attach a weight to a piece of elastic and then swing it in a circle. As the weight spins more quickly, the elastic stretches so the weight can move outward. It also explains why spinning salad in a centrifuge squashes the leaves up against the walls.) However, gravity keeps the air too snug to the surface to allow it to move far enough upward to accommodate its extra spin.

The only other option left to the air is to move to a point on the surface that is farther from the central axis of spin, the line that runs from north to south right through the middle of Earth. In other words, the air needs to find a place where the planet is naturally wider. The closer you go towards the equator, the wider Earth becomes, so eastward-moving winds turn towards the equator – to the right in the northern hemisphere and to the left in the southern.

The same principle applies for winds moving to the west relative to the surface. Now the air is spinning slightly slower than the planet beneath, which means it needs to move closer to the axis. Since the surface itself is in the way, preventing the air from burrowing in where it stands, it needs to move to a point where Earth is naturally thinner. The farther away from the equator you go, the thinner Earth gets, so westward-moving winds turn away from the equator – which means once again that they turn right in the northern hemisphere and left in the southern.

Ferrel found that this simple additional effect of living on a spinning world – the inexorable urge for northern hemisphere air to turn right and southern hemisphere air to turn left – was exactly what he needed to explain the mysterious patterns he had read about in Maury's book.

To make Ferrel's argument simpler, think about the winds in the northern hemisphere. (Exactly the same arguments will apply to the southern, but in mirror image.) Like most things in nature, air prefers to even things out – it will always want to move from high pressure to low pressure, and from hot to cold. At the equator, the sun's heating is intense, so the air there will rise vertically upwards.

This equatorial air can't go on rising for ever. Instead it reaches a natural ceiling, bends over, and begins to move north towards the cold pole. But moving air is always tugged sideways, so this northward air will begin to swerve towards the east. Because this is happening at high altitude, it goes unnoticed at the surface. However, the displaced high-altitude air leaves a low-pressure gap beneath. Surface air from farther north will naturally rush south to fill the gap; and since moving air always turns, this air will slew to the right, creating the easterly trade winds.[13]

Meanwhile, the high-altitude air is now moving eastwards. Because it is also cooling, it begins to fall again, more or less over the tropics. As it's falling, the air is still being tugged to its right, which means it curves south. This is the air that closes the loop and turns back into the surface trades.

At the other side of the tropics, more air is moving northwards towards the cold poles. But this air also has the perpetual urge to turn right. Before it can arrive at the North Pole it, too, is forced to turn to the east, and then to the south.

Now Ferrel was left with two perpetually diverted currents of air, one heading south from the North Pole and one north from the equator. These, he realised, would collide at the tropics, piling up to create that mysterious mountain range of high air pressure encircling the globe.

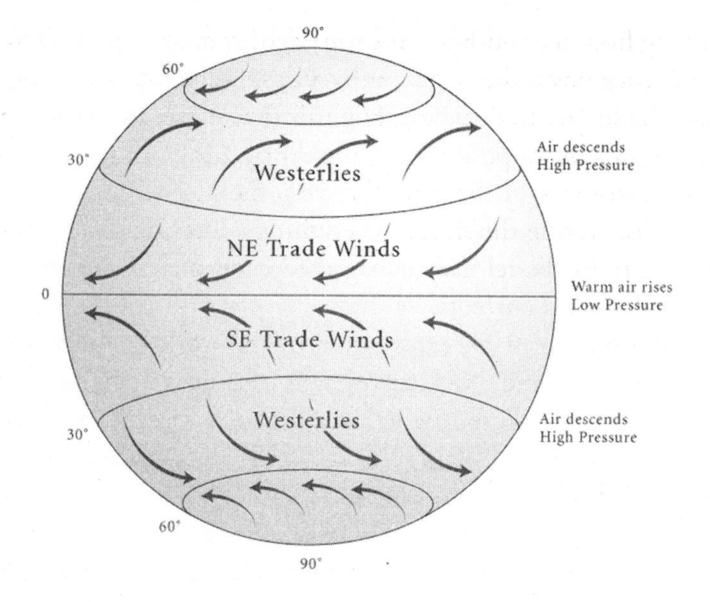


Air descends at the tropics to create two 'mountain ranges' of high pressure. Equatorwards of this, air moves to the west, while polewards it moves to the east.

And because air can only move into a part of the atmosphere where the pressure is lower, this mountain range acts as a barrier. That means the surface beneath is scarcely ever refreshed by damp air, which would bring rain. This is why Earth's deserts occur mainly in two giant rings centred about 30 degrees north and south of the equator, even though the equator itself is hotter. (You can check this on a globe. Trace the positions of the Sahara and the deserts of Asia and central America in the northern hemisphere, and the deserts of South America, Namibia, and Australia in the southern.)

Finally, the high-pressure system set up by those two currents of right-turning air explains beautifully the origin, location and persistence of the two great bands of winds that Columbus discovered: the trades and the westerlies. These two bands form on opposite sides of the high-pressure zone. This mountain range of piled-up air behaves as if it has sloping sides to the north and south. Equatorial air that has risen to great heights and is arriving from the south hits the top of the mountain and is forced to roll back down the slope southwards (and hence westwards) to form the surface trade winds. High air that arrives from the north slides down the opposite slope, then turns back northwards (and hence eastwards) to become the westerlies.

By discovering the simple 'everything turns right' rule among his equations, Ferrel had managed to explain nearly everything about the wind currents.

Ferrel published his explanation of the world's winds as 'An essay on the winds and currents of the ocean' in his friend Bowling's *Nashville Journal of Medicine and Surgery*, in 1856. It wasn't exactly the world's most widely read publication for meteorological research, but nonetheless word of this and Ferrel's other investigations began to trickle out. The following year he received an unexpected invitation to move to Cambridge, Massachusetts, to work on the *American Ephemeris and Nautical Almanac*, which was published by the US Naval Observatory. Though it wasn't an academic appointment, Ferrel eagerly accepted, and he suddenly found himself among thinkers and scientists who, much to his surprise, seemed to consider him one of their own.

From there he was poached by the US Signals Service in Washington DC, who insisted on republishing his *Nashville Journal* paper and other obscure essays so that the world's meteorologists could cast away their poor, tatty copies of the originals. Ferrel never sought a single academic position or honour, but people persisted in giving them to him. He became a member of the National Academy of Sciences, an associate fellow of the American Academy of Arts and Sciences, an honorary member of the meteorological societies of Austria, Britain and Germany, and received the honorary degrees of MA and PhD.

Ferrel would produce many other findings in his time, on an impressive variety of topics. He came up with the mathematical equivalent of labour-saving devices: formulas that made very effective shortcuts in long calculations. He worked out a new and better way to calculate pi, the ratio of the circumference to the diameter of a circle. He even calculated the weight of the moon, and explained why a throbbing star called Algol was winking increasingly rapidly at Earth. Though his scientific career didn't begin until he was nearly forty, by the time he retired thirty years later he had produced some three thousand pages of scientific research.[14]

In all that time, Ferrel never lost his paralysing shyness – a failing he was well aware of but couldn't seem to shake. Once, he wrote a paper, 'Note on the influence of the tides in causing an apparent acceleration of the moon's mean motion', and realised that it contained an original and important finding. He decided to present the paper to the American Academy of Arts and Sciences, but somehow he couldn't bring himself to stand up and read it aloud. 'I carried it to the meetings of the Academy time after time with the intention of reading it,' he later confessed, 'and my courage failed.'[15]

William Ferrel finally retired to the Midwest, age seventy, but he couldn't bear to be so far away from a ready supply of books and soon moved back east. He died five years later, as peacefully as he had lived. One of his greatest friends, the meteorologist Cleveland Abbé, who had known him for thirty years, wrote: 'We all remember his quiet ways, his indefatigable industry, his shyness, his perpetual absorption in the contemplation of some new and complex problem. He lived in an atmosphere of abstraction; he was with us, yet not of us.'

Another, more formal, obituary, by a meteorological professor who had worked a little with Ferrel, said, 'It is a curious commentary on renown to name Ferrel, of whom the great world knows nothing . . . as one of the most eminent scientific men that America has produced.'[16] America has produced plenty of eminent scientific men and women since that was written, but it still holds true on both counts. Ferrel is still relatively unknown, and he remains one of the best American scientists who has ever lived.

*

Without the great wind belts of the world, our planet would be a very different place, part frozen, part fried. If all the heat that arrived in the tropics from the sun simply remained there, the equatorial region would be a full 14 degrees Celsius warmer than today, and life there would be all but impossible. The poles would be in a worse state – they need to gain heat even more than the tropics need to lose it. Not only do the poles receive less direct sunlight than the equator, but their white caps reflect a large proportion of their sun back into space. Because of this, if they received no helping hand from farther south, both of the polar regions would be 25 degrees Celsius colder than today, and this cooling would spill over into the heavily populated middle to high latitudes. In other words, if heat stayed only where it landed, most of Earth's surface would be unbearable.[17]

Global winds are the agents that perform this redistribution of thermal wealth, and the middle latitudes are the engines of their endeavours. The most important ingredient of this engine is another vital aspect of moving air that Ferrel's new 'north turns right' rule managed to explain: not wind currents, but storms.

Until Ferrel came along, nobody had understood why the winds around storms and weather patterns move in circles. In fact, before satellites could show us all the dramatic images of spiralling hurricanes, many people refused to believe that they were circular. The explanation for the shape of storms lies very simply in the new rule that Ferrel discovered. Every piece of moving air on Earth has the perpetual urge to turn. The sky is never still, nor is it uniform. Air is always on the move from one place to the next, and storms begin with any small patch of sky that has lost a little air to its neighbours. Air from the surrounding regions starts to try to fill this hole, but it can't. As soon as it starts to move towards the low-pressure centre, it has to turn. Northern air turns right, which is why northern cyclones are always anticlockwise. Similarly, southern storms move clockwise, because the air there turns left. Every circling storm is a direct manifestation of our constantly spinning planet, and Ferrel was the first to understand that.

The energy that feeds tropical storms comes from the water they suck up from the sea, but that constant circling is why they can keep their fierce destructive energy for such long periods. Like an ice-skater pulling in her arms to spin more quickly, the air closest to the calm, low-pressure centre is the fastest. The 'eye' of a storm is usually surrounded by the strongest winds, circling desperately, unable to make that final leap into the centre.

Hurricanes, the most powerful storms in the world, are found in the lowest latitudes, where the trade winds blow. Without copious amounts of warm water to feed them they can't exist, which is why you get hurricanes only in the tropics.[18] However, they don't happen very often, and then only in the warmest season of the year. Also, they're smaller, tighter and shorter-lived than the loose-limbed storms that roll chaotically around the middle latitudes.[19]

So although hurricanes are dramatic as well as destructive, they are not the main engines of climate. Instead, most of the vital work of transferring energy between tropics and poles takes place in the middle latitudes. These are giant collision zones, where cold polar air clashes up against warm tropical air and the energy released makes storms that roll around the world like ball-bearings.

This is the region that matters for our atmosphere's distribution system. It is also the only part of the circulation system that bears Ferrel's name. Drawing together all his findings, Ferrel had pictured an atmosphere containing three giant loops in each hemisphere. The first stretched from the equator to the tropics and was as George Hadley had described it; hence it was called a Hadley cell. The third went from the middle latitudes to the poles and also worked because of direct differences in the amount of heat energy the air received, so that was also a Hadley cell. The one in the middle worked the opposite way round, and formed only in response to the other two. This is often called the Ferrel cell.

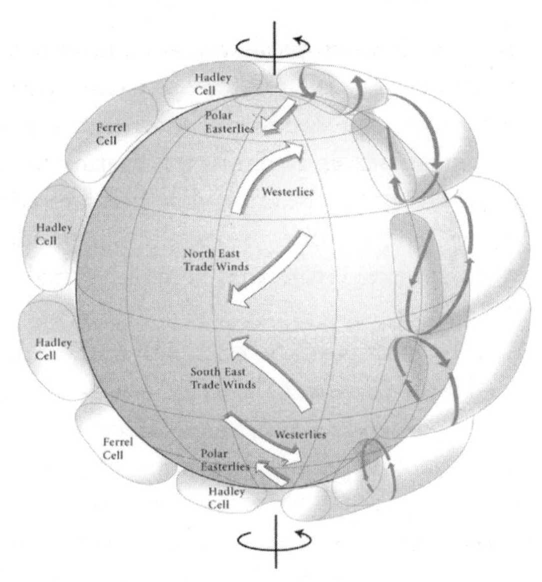


This three-cell picture of the atmosphere explains the directions of the trade winds and westerlies, though the middle 'Ferrel' cell is in fact a stormy, complex region where warm air clashes with cold to transfer heat energy from the equator to the poles.

This three-cell picture of the atmosphere is still taught in many schools, but it's not quite right. The Hadley cell near the equator certainly exists, and so does the one at the poles (though it's much weaker). But in between, there really isn't a neat cell at all; instead it's a messy confusion of whirling storms and weather systems. However, this is the meaty part of the atmospheric motions, the place that does most of the work. That's why the westerlies are so much fiercer than the trades, and why Columbus encountered that Valentine's Day storm that nearly cost him his life; they occur in the exact region where stormy weather fronts are whipping up their climatic energy and then passing it on.

It's also why mid-latitude weather is so unpredictable. If you live in Hawaii, in the path of the trade winds, gentle breezes will blow from the east almost every day of the year. But for those of us who live in middle latitudes, right underneath Ferrel's rolling storms and circling weather fronts, the weather can come from almost anywhere. (Mark Twain put this beautifully, in a forecast he made for a typical New England day: 'Probably nor'east to sou'west winds, varying to the southard and westard and eastard, and points between; high and low barometer, sweeping round from place to place; probable areas of rain, snow, hail and drought, succeeded or preceded by earthquakes with thunder and lightning.')[20]

But there was a missing piece to this puzzle, one that even the great Ferrel didn't see. He didn't know it, but his two major explanations, the presence of the westerlies and the spinning of storms, were intimately related.

Ferrel had thought the surface westerlies were probably balanced by similar-strength but opposite 'easterlies' aloft. In this, he was wrong. The westerlies are merely the trailing skirts of a more dramatic and much more violent wind whose existence nobody on Earth suspected. Even a sky that seems blue and empty sometimes holds a river of air even stronger than a hurricane. This is the agency that guides the spinning cyclones of the middle latitudes with a high invisible hand, and is the final ingredient for how air spreads out its life-giving heat and rain.

**19 July 1933**

Alone in the cockpit of the *Winnie Mae*, Wiley Post contemplated the dense bank of cloud below him. He'd always known that Siberia would be the most dangerous part of his attempt to fly solo around the world. Its high peaks were usually hidden in fog, and the wildly inaccurate Soviet maps were of no help. Wiley had learned that lesson two years earlier, when they all but led him into the side of a Siberian mountain. That was on his first round-the-world flight, when he'd had a navigator helping him. This time he was flying alone and half-blind. (A white cotton patch was tied, as usual, over the socket where his left eye should have been.)

He also knew that there were plenty of people who would be perfectly happy if he didn't return. Post may have been a genius at flying – someone once said of him that 'he doesn't fly the plane, he wears it' – but that was nothing compared with his talent for making enemies. Whether because of his poor background, his short stature, or just the kind of personality that assumes the world is out to get you, Post complained at and harassed and flew into tempers with even the people who were trying to help him. He had so antagonised the press that one cameraman reportedly told him he wasn't out on the runway to get pictures of the take-off. Rather, he was hoping that Post's plane would crack up and he could 'get a picture of you frying on both sides'.[21] Apparently, he wasn't joking.

Fortunately, Post had learned another lesson from his previous Siberian flight. To avoid trouble, he could simply fly high. You weren't supposed to be able to fly at 20,000 feet without oxygen, but Post had discovered he could manage it if he didn't stay too long. Besides, there was something else he wanted to check. Back in 1931 on the first round-the-world trip, when he and his navigator had taken their craft up into the clear blue sky above those dangerous clouds, they had encountered a sudden tailwind so strong that their speed had increased by 100 miles per hour.

Post was convinced that the aeroplane would be the transport of the future, and that long-distance flights would be the key to their success. Why bother spending five days on a ship to Europe if you could be there in only a few hours? However, there were still many sceptics, for whom planes remained little more than a curiosity, and he had become determined to prove them wrong. If any extra winds could boost his efforts, he wanted to know more.

As he floated above those clouds in Siberia, Post found what he was looking for. He felt that same sudden turbulence and then the unexpected boost from a roaring river of air, even though the sky around him was apparently serene. A few days later he felt the boost again as he crept to 20,000 feet over Alaska. There really did seem to be something up there, something a pilot could use.

When Post landed in New York, he had broken his previous record by more than twenty-one hours. The city gave him a ticker-tape parade. Everybody wanted to use his name for endorsements. Camel cigarettes enlisted him as a 'famous smoker', even though he didn't actually smoke. Their ad contained a purported quote from Post: 'It takes healthy nerves to fly around the world alone. Smoking Camels as I have for so long, I never worry about healthy nerves.'[22] Sometimes he'd even pretend to smoke a Camel, before stubbing it out after a couple of drags.

But when he started babbling about how a sky that was obviously clear and empty was instead full of hurricane-force winds, there were plenty of people ready to laugh. They attributed his declaration variously to sitting too long in one position, lack of oxygen or just plain flights of fancy as he sat there in his plane, on his own, with nobody to talk to, nothing else to do . . .

Post couldn't bear to be doubted. He had come a long way for this, and he was determined to prove he was right.

Like William Ferrel, Wiley Post began as a poor farm-boy. But his was a very different character, born into a very different age. In Ferrel's time, America was still a country of pioneers and struggle. But by the year of Wiley's birth, at the other end of the nineteenth century, the world was changing so quickly that news of it reached even his rural community of Corinth, Texas. In 1898, cities had experienced such a spurt of growth that New York contained a staggering 3.5 million inhabitants. And near the cities, large industrial 'manufactories' were springing up to make their centralised products and spread them throughout the country. For instance, a new breakfast cereal called 'cornflakes' had started appearing in the shops. (It was an unhappy beginning since the formulation wasn't quite right, and the flakes turned rancid on the grocers' shelves.)[23] In the same year, nearly one thousand of the new horseless automobiles rolled off American production lines. And five years later, at Kitty Hawk in North Carolina, the Wright brothers would prove that men could fly.

Wiley Post saw his first aeroplane at the age of fourteen. Unlike Ferrel, Post was no scholar. Sublimely uninterested in learning, he had quit school three years earlier and even now could barely spell his own name. But he'd always loved technology and he had an uncanny knack with machines. The young Post could fix just about anything, from sewing machines to seeders, and had spent his time since leaving school going from farm to neighbouring farm offering his services. The year before, he had earned enough money to buy one of America's first bicycles.

But he forgot all about bicycles in the autumn of 1913, when he went with his family to the county fair in Lawton, Oklahoma. Art Smith, one of America's first barnstormers, had brought along one of the first aeroplanes: a Curtis Pusher. Post ignored the amusement rides, the exhibitions, the displays of harvesters and planters. All that mattered was the plane. 'I have never seen a bit of machinery for land, sea, or sky that has taken my breath away as did that old Pusher,' Post later said. When the fair was over, and it was long past their prearranged meeting time, his brothers found him sitting in the Pusher's empty pilot seat, his head full of dreams.

From that day onwards, all Post wanted was to fly. But planes were new and exotic, and above all expensive. Nobody was going to hire an inexperienced farm-boy to fly a plane, especially one who looked like Post. He was short, only five feet four inches tall, but he was also tough, with a belligerent temper that matched his flaming hair. What Post really needed was the money to buy his own plane. And if finding it involved a little risk-taking, so much the better.

So Post became a highway robber. The scheme was ingenious, and much practised in the later Depression: leave an apparently abandoned piece of machinery in the road to tempt passers-by to stop, and then leap out, brandish a gun and demand their money. However, Post can't have been very good at it – he had apparently been terrorising Grady County, Oklahoma, for some time when he was arrested in 1921, but he still had only twenty-seven cents in his pocket. Though he was sentenced to ten years in prison, he became so depressed and withdrawn that he was paroled after only a year. Post was always deeply ashamed of his criminal record and went to great lengths, when famous, to hide it from the world.

The terms of Post's parole meant that he had to keep his record very clean. But that didn't mean he had to give up his dream of flying. He started working in the oil fields, but it wasn't long before he had found his way to Burrell Tibbs' Flying Circus and offered his services. He was lucky. As it happened, one of the circus's most important acts, the parachutist, had been injured the day before. The crowds loved watching a man leap to his possible death, and the circus urgently needed a replacement. Had Wiley ever jumped before? No? Well, not to worry; there's nothing to it. All you really have to do is strap on the chute and jump. There turned out to be rather more to it than that, but when Post finally remembered to pull the release cord, he floated down to Earth with a sensation that he called 'one of the biggest thrills of my life'.

More jumps followed, along with some flying lessons and a little barnstorming. But even at fifty dollars a jump, Post couldn't earn the money he'd need to buy his own plane. Reluctantly, he headed back to the oil fields, to a drilling company named Droppleman & Cunliffe. His first day at work was 1 October 1926. It was also his last. The details of the day, like many of the stories surrounding Post, are slightly murky. A rotary chain had broken, and someone had taken up a hammer to knock a pin out of the chain. Post always claimed that someone else had struck the blow, but one of his co-workers later swore that it had been Post himself who inadvertently caused his own injury. In any case, what is certain is that a piece of steel broke off and flew into Post's left eyeball. Though a doctor tried to work the steel out with a needle, infection set in and a specialist was forced to remove Post's eye. Thirty days later, the State Industrial Commission of the State of Oklahoma ruled that Post was entitled to $1,698.25 compensation for his injury.

Post may have lost an eye, but now he could have his plane.

But first, Post had to learn to compensate for his injury. This was long before the days of reliable instruments to gauge a plane's height from the ground and distance from approaching obstacles, and with only one eye, he would struggle to judge such distances for himself. During his recuperation, Post spent several hours a day going for long walks. He would estimate the distance to that rock, or that cliff or mountain, and then pace it out to see if he was right, or the height of a tree and then climb it to check. His guesses became more and more accurate until, finally, Post decided that he was ready.

Post's new plane, the *Jenny*, was a Canuck, built in Canada. She swallowed up most of his compensation money and he spent much of the rest on flying lessons. The plane turned out to be most convenient, and not only for business purposes. Post hired himself out as a pilot, but he also used the *Jenny* for his elopement. For Post had tumbled in love with his first cousin, Edna 'Mae' Laine, and her parents were implacably opposed to the match. Their reluctance was fairly understandable. Mae was an innocent seventeen-year-old and Post a worldly man of twenty-eight. He had only one eye, was irascible, and there was that regrettable history of highway robbery. And bouncing around America on newfangled aeroplanes wasn't exactly a steady job. Post realised he would have to take Mae in a way that her parents couldn't stop.

On 27 June 1927, Mae and Wiley climbed into the *Jenny*. Mae had a small bag with a few possessions; Wiley had a marriage licence in his pocket. Nobody has recorded what Mae's parents shouted when they heard the plane's engine fire up and realised she was inside. Perhaps the couple waved. But their escape was still not over. Only thirty miles south of Mae's parental home, the Canuck's engine stopped. Post looked around hastily and found a field that was fairly flat, and recently harvested, where he could land. But the couple were now in the middle of nowhere, and there was no way he could get the plane fixed that night. In the spirit of respectability, not to mention the worry that an irate father might arrive any minute and ruin the proceedings, Post took off in search of a preacher, who obligingly married the couple on the spot. They spent their wedding night in the open air on the wooden platform of an oil derrick.

In the end, Mae's parents forgave Post. He certainly became more rich and famous than they had ever dreamed he would (though never quite as rich as he himself hoped). Working as the private pilot for a wealthy businessman, Post first won the National Air Race Derby from Los Angeles to Chicago. Then, in June 1931, came the first of his records, when he flew around the world with his navigator, Harold Gatty; the pair had travelled 15,474 miles in eight days, fifteen hours, and fifty-one minutes. During the entire trip, Post had slept less than fifteen hours.

After that, and the solo record that followed, Post had all the fame that he could ever have wanted. But what he wanted even more now was to be believed. He had felt those high-altitude winds. He knew they were there. And he would prove it.

To do this, he was going to have to go high and stay high, and for that he would need oxygen. The *Winnie Mae* was far too leaky to be pressurised, but perhaps he could have the oxygen piped into his own suit. Now there was a mechanical challenge fit for Wiley Post. He began furiously designing one suit after another. The first was a two-piece, joined by an airtight belt, with pigskin gloves and rubber boots, and an aluminium helmet like a welder's with a trapdoor over the mouth so he could eat or drink. Sadly, as soon as he tried it on in a reduced-pressure chamber it unmistakably leaked. The second attempt was a little better, but although Post could get into it easily, he then found himself embarrassingly stuck. He had gained twenty pounds since his body measurements had been taken for manufacturing the suit and after several futile attempts at removal, it had to be cut off.

The third time, though, was the charm. To guarantee the suit would fit, Post had an exact metal replica made of himself sitting comfortably in a chair, just as he would in the cockpit. This was then covered with latex to form the suit's inner shell. Oxygen came into the helmet from the left side, near his missing eye, so the flow of air wouldn't disturb his vision. The tests looked good. Post was ready to try what would prove to be the world's first spacesuit.

First, Post simply wanted to fly as high as possible and make sure the suit held. On 5 September 1934, he reached an altitude of 40,000 feet and the suit worked well. On 7 December he took off again and made it to 50,000 feet, a new record for powered flight (though the record remained unofficial, since one of Post's two barographs had frozen at 35,000 feet). Both times, Post felt the unmistakable shove from that high-altitude river of air. 'As a result of this flight, I am convinced that airplanes can travel at terrific speeds above 30,000 feet, by getting into the prevailing wind channel,'[24] he reported.

And now came the real test. Could he use the high winds to fly faster than the plane should conceivably be capable of flying, fast enough that his doubters would have to believe him? His first attempt came very close to disaster.

On 22 February 1935, Post took off from Burbank airport in California, bound for the East Coast. He climbed almost straight up, five miles high. But then, at 24,500 feet, he noticed a problem. The oil pressure suddenly dropped. If he didn't cut off the engine, every bearing in the plane would jam. Post began to drop in altitude, looking for somewhere to land the plane. He was only thirty-five minutes out of Burbank, but there was nothing like a runway in sight. Worse, he had jettisoned his landing gear to make the plane more streamlined. He would have to land on the plane's specially reinforced fuselage, and it was surely going to be rough. But then Post spotted a dry lake bed and, with true flying genius, managed to bring the plane down and to a safe halt.

He struggled his way out of the cockpit, but there was no way he could remove the pressure suit on his own. Hampered by the thick cloth, Post couldn't even reach around the back to unscrew his helmet. In the end he walked to a road, where a motorist was tinkering with his broken-down car. It must have been quite a sight. 'The man's knees buckled and he almost fell over. He ran around to the back of his auto and peered at me. I had a time calming him down but I finally succeeded and he helped me out of my oxygen helmet. "Gosh fellow," he exclaimed when he found his voice, "I was frightened stiff. I thought you had dropped out of the moon, or somewhere." '[25]

The two men walked together to get help. But it was only when the plane was brought back to Burbank for inspection that Post discovered what had gone wrong. Two pounds of metal filings and dust had deliberately been poured into the oil tank. Someone had tried to kill him.

Wiley Post, daredevil, ex-parachute jumper and highway robber, wasn't the sort to scare easily. On 5 March 1935, he climbed back into a plane that had been meticulously checked and set off again from Burbank. This time, all went well – at least at first. But when he reached Ohio, Post realised that he was almost out of oxygen. He had no choice but to drop back down to lower altitudes and land at Cleveland airport. Still, he had surely done enough. The *Winnie Mae* had flown 2,000 miles in seven hours, nineteen minutes. That made her average speed nearly 280 miles an hour, which was at least 100 miles an hour faster than she should be able to fly.

But somehow it still didn't convince people. Post was simply too unreliable a witness to overcome their prejudice against the idea of high winds in clear air. Perhaps if he could just get all the way to the East Coast . . . But though he tried again, twice, mechanical failure kept bringing him down. According to one report, in the first of these attempts his helmet fogged up so much he had to clear it by scraping it with the tip of his nose. When his nose was so raw that it had smeared the glass with blood, he had to land.

It's a pity that nobody believed Wiley Post. He would die in a flying accident in Alaska (apparently an innocent one) years before the world realised he had been right. And in the meantime, human ignorance of his discovery would lead to more than one tragedy.

What Wiley Post had called 'high winds' we now know as jet streams. These fast-flowing rivers of air circle the world in both hemispheres. They're not always invisible. Sometimes they drag white cirrus clouds along for the ride, and the long, thin trail they leave can be seen from space. Mount Everest pokes up into one stream that blows from west to east over Asia, which is why many portraits of the mountain show a trail of blowing snow over the eastern face. Jet streams are only a few hundred kilometres across and perhaps only a few kilometres deep, but they are fierce. Whipping along at speeds of more than 150 kilometres and sometimes as much as 480 kilometres per hour, they are among the strongest winds in the world: faster than hurricanes, almost as fast as tornadoes, but with an influence that is much more far-reaching.

Post had caught his own glimpse of the jets over Siberia and Alaska, but the next time they showed up was above Japan, towards the end of the Second World War. American B-29 bombers had been specially designed to fly above 30,000 feet, so they could evade enemy fighter planes while preserving their bombing accuracy. But, bizarrely, when they arrived over Japan their targeting went completely awry. The bombers should have been travelling at 340 mph but their instruments claimed they had a ground speed of 480. At that speed there was no way they could zero in on targets five miles below. The commanders were more inclined to blame the pilots for incompetence than to credit their tales of hurricane-strength winds above the clouds, but doubts were beginning to set in.

Still, nobody knew that these rivers of air could be anything more than a weird local effect until, on a hunch, the Japanese military released thousands of booby-trapped balloons in early 1945. The balloons were equipped with an ingenious device to keep them floating in the jet stream: if they dropped too low, a pressure sensor would detonate a charge and a small bag of ballast would be jettisoned into the Pacific. The military had no idea how far the balloons would get, but one thousand of them made it all the way to the west coast of America. By hitching a ride on their invisible river, they had travelled 6,000 miles in only four days.

Many were shot down. Some were captured. 'Japan attacks US mainland with bomber balloons' was a headline guaranteed to cause panic, and the US military was determined to keep the story out of the press. Then, on 5 May 1945 in Bligh, Oregon, a group of Sunday school children went out to the woods for a picnic. It was a beautiful summer day, and the kids suspected nothing as they raced over to the strange device that lay in a clearing. Nobody knows who touched it first, but their bones were embedded, along with shrapnel, into the surrounding trees. Five children and their teacher were killed. They were the only fatalities on mainland America during the entire war.

But while people in the northern hemisphere were discovering the power of the jet streams, nobody yet suspected that they might also occur in the south. Commercial flight was just beginning, and few planes could fly high enough to notice the change in wind. One of these few was a British Lancaster airliner called *Stardust*, which had been designed to fly high over the Andes in case it needed to avoid the storms and clouds that often hugged the mountain peaks. On 2 August 1947, *Stardust* took off on a straightforward flight from Buenos Aires, Argentina, bound for Santiago, Chile. This would involve a simple hop over Mount Tupangato, one of the highest peaks in the Andes.

According to the weather reports, visibility would be poor, so as they approached the mountain the *Stardust*'s pilot radioed his intention to climb up over 24,000 feet. Radio contact continued as normal; the pilot reported that he had crossed the mountain and was about to descend into Santiago airport. Then, without warning, the plane vanished.

After fifty baffling years, investigators have finally worked out what happened to the *Stardust.* It wasn't alien abduction, a South American 'Bermuda Triangle' or any of the other bizarre theories that had been advanced in the intervening time. Instead, the unfortunate plane had encountered the southern jet stream. As *Stardust* rose to 24,000 feet, she suddenly encountered a fierce headwind blowing her backward at more than 100 miles per hour. The problem was that the pilot didn't know this. He had no radar to tell him that his ground speed had just dropped by nearly half, nor were there any radio stations tracking his position from the cloud-covered, uninhabited ground below.

All he could do was calculate his position according to the speeds the instruments gave him. So when he thought he had arrived near Santiago, he had not yet cleared the mountain. When the plane crashed into the eastern face of Mount Tupangato, the three crew members and six passengers on board were killed instantly. Seconds later, an avalanche triggered by the impact covered the plane in a blanket of snow, from where it gradually sank into the heart of the glacier and made its frigid way down to the valley floor.

If the glacier hadn't spat out the remains of the plane some fifty years later, nobody would have known that the jet stream had claimed more victims. And yet, the streams themselves are far from malevolent. Now that we understand them and can monitor where they form, they are even living up to Wiley Post's dream of using their power. A boost from the powerful jet stream that blows east from North America to Europe explains why transatlantic flights are nearly an hour faster going east than going west. And in 1999 a balloon, the *Breitling Orbiter*, used the jet streams to hitch a non-stop ride around the world.

But the jet streams are much more important than this – they are the final step in making our planet habitable. For they act as guides for the rolling ball-bearings of Ferrel's circular storms.

Jet streams tend to occur at either side of Ferrel's stormy westerlies, in the high atmosphere; where tropical air meets the cooler air of the middle latitudes; and again where middle-latitude air collides with the freezing atmosphere of the polar regions. Take the northern hemisphere jet streams. In each case, the big contrast in temperature between these two bodies of air sends the southern air roaring northward, which according to Ferrel's 'north turns right' rule means that it whips around to the east. Each of these two jets thrashes about in complex ways. Sometimes the two of them merge to make one gigantic jet in each hemisphere; sometimes, they all but disappear. They are strongest in the winter, when the temperature contrast between equator and pole is the greatest.

Storms form in the same regions as jet streams because they, too, feed off the strong temperature contrasts, and the jets then steer them around the world. The rain contained in these storms is one of the chief engines of climate, the means by which our air can redistribute its resources – taking from each parcel of air according to its ability to give, giving to each according to its needs.

Though our atmosphere contains only a few hundredths of a per cent of all the water on Earth, it is by far the most active transporter.[26] An average molecule of water will stay locked in oceans and ice sheets for hundreds or thousands of years, but one that is soaked up into the atmosphere will be carried aloft and then rained out again in only ten days.

Life could not survive anywhere on Earth's land surface without rainfall, for all living things need water, and without the atmosphere's helpful redistribution mechanisms we would be confined to the seas. But more than that, the storms that bring water also carry heat.

When air soaks up water from the ocean, it uses energy to rip the water molecules apart from one another and turn them into a disintegrated gas. When the molecules reunite to form raindrops they give out energy, which is what feeds the storms. Heat and water are intimately connected, and the global winds redistribute both. (The same principle is behind sweating. When you sweat, glands take water from the interior of your body and pour it out on to the surface. This water then gradually evaporates into the air around you, taking with it your excess heat energy, to be delivered via air and rain, to somewhere else that needs it more.)

Earth's gigantic wind systems have been performing this feat for billions of years, producing many different patterns of global climate. The winds adapt to subtle shifts in the gradients of temperature and the amounts of available water to produce worlds that have always been habitable but have sometimes looked quite different from the one we have today. However, we humans have evolved in a world with one specific set of handouts, and one specific resulting climate. And our own particular pattern of redistribution may soon change. Many now fear that global warming will interfere with the way the winds deposit their loads. Warmer air can hold more water before it must be shed as rain, perhaps bringing droughts to some regions. More water in the air means more energy, so storms may be fiercer. As the polar regions warm, jet streams may shift their positions; some think that the widespread fires in North America in 2002 were a symptom of the jet stream shifting north and taking its rainstorms with it.

But even if all this does take place, Earth will probably adapt. There are still likely to be lakes and rivers and reservoirs at least somewhere on the planet. Our enveloping air has effected this transformation of the Earth for more than four billion years, and there's no reason it should stop now. (Whether the adapted Earth will still be a comfortable, or even feasible, place for large numbers of humans to live is quite another matter.)

So far, we have seen our ocean of air all in the guise of transformer. But it has another role, just as crucial for the survival of Earth's creatures. For the life the air engenders is still vulnerable. Space is filled with hazards that, if they ever reached the planet's surface, would put us all in grave peril.

Here again, our atmosphere intercedes for us. Above the clouds, layer after layer of air provide bulwarks against the ravages of space. And the very first of those protective layers was nearly destroyed almost before we knew it was there.

# PART TWO

# SHELTERING SKY

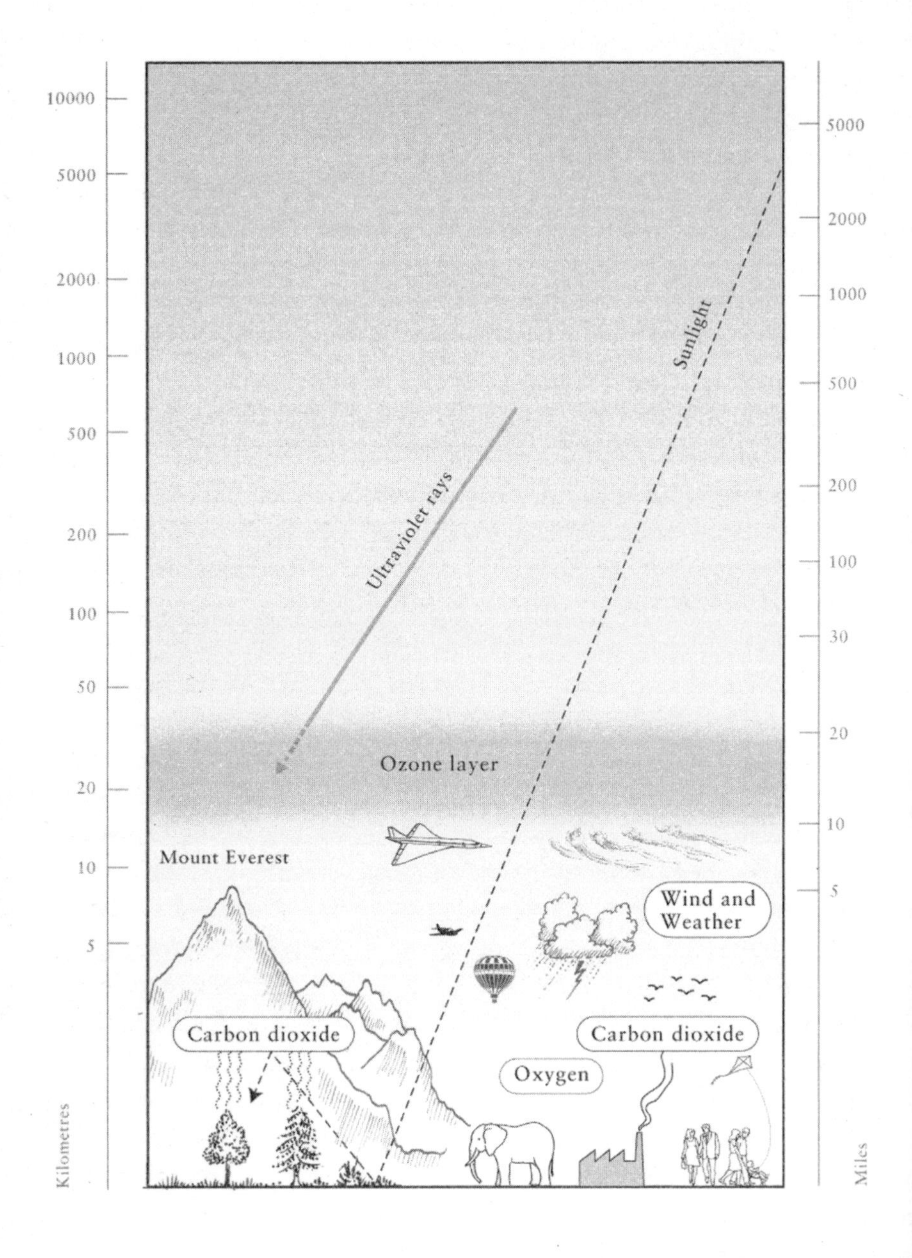

10000
5000
2000
1000
500
200
100
50
20
10
5
Kilometres
5000
2000
1000
500
200
100
30
20
10
5
Miles
Sunlight
Ultraviolet rays
Ozone layer
Mount Everest
Wind and Weather
Carbon dioxide
Carbon dioxide
Oxygen

*Chapter 5*

# THE HOLE STORY

Ozone is a beautiful gas. Unlike its closest relative, oxygen, which is invisible, ozone is a vibrant shade of blue. When Dublin scientist W. M. Hartley began working on the gas in 1881, he was enchanted by its colour, 'as blue as the sky on a brilliant day'. And though some people were inclined to find the smell of ozone disagreeably pungent, Hartley thought it fresh, as after a great thunderstorm when the world has been washed clean: 'Ozonised air gave a very distinct odour, quite unmistakable, but quite reminding one of the air on the South Downs during a south-west breeze.'[1]

Hartley was curious about this new gas, discovered only forty years earlier. It existed naturally in the environment, but apparently only in tiny quantities and special circumstances, such as after a lightning strike. Some researchers had recently discovered that it was made of oxygen atoms; but where the molecules of normal oxygen contain only a pair of atoms ($O_2$), ozone molecules have a third ($O_3$). This additional atom seemed to make ozone even more reactive than oxygen. Breathing it was an uncomfortable experience. It caused chest pains and irritation, and small animals such as mice couldn't survive in it for long. (At ground level in the modern world, ozone is a component of automobile smog, and hence a major irritant for asthmatics.)

But that wasn't the whole story. Hartley was about to discover that high in the atmosphere, ozone plays a very different part in our lives. Starting some thirty kilometres above the ground, it forms a protective layer, the first of the air's three silver linings that shield every living creature from the hostility of space.

He was led to this discovery by the curious observation that some of the sun's rays were missing. Recall that the sun throws out more kinds of light than we humans can see. Beyond the red end of the rainbow lie the long infrared light waves that are responsible for warming our planet. Their successive peaks and troughs are too widely spaced to be seen by our limited eyes. But infrared light also has a high-energy cousin called ultraviolet, which appears beyond the blue end of the rainbow and whose waves are too short for us to see.

Though our eyes are blind to these extra rays, by Hartley's time there were plenty of instruments that could spot them. And there lay the problem. The infrared rays were there all right, but the ultraviolet ones suddenly stopped. Visible light cuts off at a wavelength of around 400 nanometres (which is four ten-thousandths of a metre). Anything shorter than that is ultraviolet light, and you'd expect ultraviolet rays from the sun at every wavelength from 400 all the way down to 200 nanometres. But below 293 nanometres, there was nothing. Or at least nothing that arrived at Earth's surface.

Either the sun wasn't putting out these highest-energy, shortest-wavelength ultraviolet rays, or something was stopping them from reaching us.

Hartley was thinking about this problem when he noticed that ozone gas had a tendency to absorb ultraviolet rays. What, he wondered, would happen if he tried shining a full complement of ultraviolet wavelengths through a bright blue tube of ozone gas? The answer was that the ozone clipped off the end of the ultraviolet rainbow. Nothing shorter than 293 nanometres made it through to the other side. Hartley concluded his paper describing these experiments with the following words, written in what was for him an unusually formal manner:

> The foregoing experiments and considerations have led me to the following conclusions:
> 1st That ozone is a normal constituent of the higher atmosphere
> 2nd That it is in larger proportion there than near the earth's surface
> 3rd That the quantity of atmospheric ozone is quite sufficient to account for the limitation of the solar spectrum in the ultraviolet region.

He was right. Five billion tons of ozone float above our heads, trapping the highest-energy ultraviolet rays before they make it down to the surface. The lowest-energy ultraviolet rays, the ones that our ozone frontier-guards let through, are quite good for humans. They encourage our skin to make vitamin D, which we need to avoid rickets and other bone diseases, and they also give some of us golden brown tans. But if the ones it trapped were allowed to fall freely to ground, they would be highly dangerous. These forms of UV light, called UVB and UVC, attack whatever they touch. They weaken the human immune system; they cause skin cancer and cataracts; and they destroy algae, which are the most fundamental members of the ocean food-chain.

Our ozone layer protects us so comfortably and effectively that we could easily never know the dangers that lie just a few miles above us. It works like a minefield: whenever an ozone molecule is touched by an ultraviolet ray, it explodes, firing off one of its three oxygen atoms. But this is a minefield that reforms itself constantly. The shrapnel from the explosion – a stray oxygen atom and an ordinary oxygen molecule – recombine. And when they do, the ozone is born again.

This much was figured out by a British chemist named Sidney Chapman in the 1930s, fifty years after Hartley's discovery. But at the very time that he was writing down the equations showing how powerful and vital the ozone layer is, another chemist was creating a chemical that would come close to destroying it. For, like many things that are strong, the ozone layer is also vulnerable.

In 1920s America, a certain industrialist was preparing to make another of his inventions. Thomas Midgley was a jovial man, full of enthusiasm and energy. He had hordes of friends, and – amazingly, given his many successes – scarcely any enemies. His face, round like a full moon, beamed with *bonhomie*, especially when he found a new engineering puzzle to solve. Even in his spare time he was captivated by mechanical problems. When he was walking in the countryside, he spent half his time supine, trying to figure out the principles behind the construction of anthills. When he took up golf and discovered the poor quality of the greens, he began experimenting at home with new kinds of grasses. He was a born inventor.

It's not surprising that Midgley had an inventor's eye – his whole family loved to experiment. His mother's father had devised the circular saw, and his own father held a number of patents for new kinds of tyres and bicycle wheels. Midgley's first job was in 'Inventions Department No. 3' of the National Cash Register Company in Dayton, Ohio, and then in 1916 he moved to the research division of the General Motors Company.[2] It was there that he would make his most famous inventions, materials that would prove to be useful, powerful and ultimately deadly.

For though he didn't know it at the time, Midgley was destined to be terribly unlucky with his inventions. One of the first things he did at General Motors was to recommend putting lead in petrol. He had a good reason for this – indeed, it was hailed as an ingenious solution to a most annoying problem. Cars and planes were relatively recent inventions, and all attempts to make their engines more efficient came up against the same problem: uneven combustion meant that they made an infuriating knocking sound and operated poorly. Midgley wanted to find something that he could add to the petrol to make it burn more evenly. At first, he had little success. He tried everything 'from melted butter and camphor to ethyl acetate and aluminum chloride . . . and most of them had no more effect than spitting in the Great Lakes'.[3] (He did discover that compounds containing tellurium and selenium seemed to work, but they had a bizarre side-effect, making the workers reek with the smell of garlic.)[4]

Finally, in December 1921, after working his way through thousands of compounds, Midgley discovered that adding lead to the mix solved everything. He had to overcome a certain amount of prejudice from consumers who thought that lead in petrol might be dangerous. (It is. It accumulates in humans and causes several debilitating diseases, which is why it is now banned.) But at the time, Midgley's well-meaning arguments prevailed. The first leaded petrol went on sale in 1923, and it quickly became universal. Engines in cars and planes could now work much more efficiently, and Midgley was on his way to becoming a hero.

Midgley's next invention arose from a problem that came to him from Frigidaire, the refrigeration division of General Motors. Mechanical refrigeration was a recent arrival on the technological scene.[5] Before then, ice had to be shipped down from Canada to provide a coolant of sorts, but it was expensive, weather-dependent and not widely available. Hospital wards in the southern United States were often unbearable during the summer, with the heat killing off as many people as actual illnesses. Food spoiled rapidly, and 'tropical' diseases like yellow fever and malaria were still rampant there. So mechanical refrigeration seemed like a miracle. Buildings could be air-conditioned, families could keep food for as long as they wanted, and people could make their own ice even in midsummer.

Refrigerators work by successively liquefying and re-evaporating the material inside their pipes. The material starts off as a gas, but in the pipes outside the fridge the gas is squeezed until it turns into a liquid – which releases heat energy and explains why the backs of refrigerators get hot. This liquid is then carried inside the fridge, where it is allowed to expand until it turns back into a gas. This process is the exact opposite of the liquefaction. It soaks up heat energy from its surroundings, cooling down the fridge in the process.

The problem lay in the choice of material that could be so readily squeezed into a liquid and then sprayed back into a gas. To date, every refrigerant that anyone tried had some kind of health hazard attached – some were toxic, some were flammable, and some were both. As long as these gases stayed safely in their closed pipes, that wasn't a problem. But somewhere, sometime, there would be a leak, and that's where the trouble started. By 1929 Frigidaire had sold one million domestic refrigerators, and the accident toll was mounting. People moved their fridges out on to the back porch. After a fatal leak in a Cleveland medical centre, hospitals scarcely dared use them at all. Frigidaire's engineers even suggested returning exclusively to the first refrigerant they had tried: sulphur dioxide. Yes, it was highly poisonous, but at least its irritating choking smell gave immediate warning of danger.

Midgley's task was to find a cure for these ills. He needed a refrigerant that was non-flammable and non-poisonous. It had to be absolutely safe.

Thomas Midgley set about this assignment with his usual dedication. He began by imagining various chemicals and calculating their likely properties, 'plotting boiling points, hunting toxicity data, corrections; slide rules and log paper, eraser dirt and pencil shavings, all the rest of the paraphernalia that take the place of tea leaves and crystal spheres in the life of the scientific clairvoyant'.[6] In the end, Midgley came up with a compound that seemed absolutely perfect. It had the right boiling point; it didn't burn; in fact, if his calculations were right, nothing would ruffle its chemical calm.

All that remained was to make sure this new chemical wouldn't be poisonous. And here, in one of the many ironies that dogged Midgley's inventions, he came very close to abandoning the whole project. To prepare his new chemical, Midgley had bought five small bottles of antimony trifluoride. He reached for one of these bottles at random and made a few grams of what would later be dubbed Freon. Then he placed it in a glass jar with a guinea pig. He watched and waited to see how the animal reacted to breathing the new gas. The guinea pig was completely unconcerned. It seemed that his new chemical was non-toxic, as Midgley had predicted.

But just to be sure, Midgley made another batch, starting with a second bottle of antimony trifluoride. This time, the guinea pig died instantly. Midgley was confused. Why should his Freon be poisonous to one animal and not the other? Cautiously, he sniffed at a third bottle of antimony trifluoride. What he smelled there was unmistakable: phosgene, the killer gas of the Great War. Midgley discovered that four out of his five bottles of antimony trifluoride had contained this fatal impurity. It was just luck that, for his first try, he had used the only pure sample in his batch. If he had picked up one of the others, and the first guinea pig had died, would he have continued with his research? Or would he have abandoned Freon for something else that would ultimately prove less deadly to the atmosphere?

'The chances were four to one against us,' he said later, 'and I often wonder if the sudden decease of our first guinea pig would not have so completely shaken our confident expectation that our new compound could not possibly be toxic, that – well, I still wonder if we would have been smart enough to have continued the investigation. Even if we had, the chances were still three to one against our using the one pure sample. I still wonder.'[7]

Now, using only pure samples, Midgley confirmed his first positive experiment. From then onwards, all the guinea pigs were fine. Freon had no apparent impact on man or beast. It was every bit as inert as his calculations had suggested. It was, in other words, completely 'safe'.

Midgley announced his invention at a meeting of the American Chemical Society in Atlanta in April 1930. He demonstrated the safety of his new gas with irrepressible showmanship: in front of a rapt crowd of chemists he took in a deep breath of Freon, and then slowly exhaled it over a lit candle. The candle went out.

Freon wasn't only non-flammable and non-toxic; it was also heavier than air. Salesmen liked to demonstrate this by pouring Freon down a staircase that had a lit candle on each step. Though the gas was invisible, you could track its progress as it extinguished one candle after another.

Midgley's new chemical was an immediate hit. Together with its family of related chemicals (known collectively as CFCs, or chlorofluorocarbons, so named because they contain chlorine, fluorine, and carbon), it quickly became America's refrigerant of choice. Because it was so 'safe', Midgley's company agreed to sell it to all of their competitors, and soon it was universal in refrigerators throughout the land.

With the Second World War came a new use for Freon. Soldiers in the Pacific jungles were being struck down with insect-borne diseases, so the US Department of Agriculture invented the 'bug bomb', a portable insecticide dispenser that used Freon as a propellant to spray the insecticide exactly where it was needed. This was the origin of aerosol spray-cans, in which everything from deodorant to hairspray could be neatly, precisely delivered with the help of Midgley's Freon. Next on the scene was the dry-cleaning industry. Then CFCs turned out to be perfect for making foam rubber for furniture. It must have seemed that Freon was a chemical panacea.

And yet, by inventing his 'safe' new chemical and offering it to the world, Midgley had created a monster.

At first nobody had any inkling of this. During his lifetime, Midgley was fêted for his achievements. He received almost every major prize in chemistry as well as dozens of other awards and honorary degrees. (He didn't get the Nobel prize, though his work would later inspire one.) The citation for one such award, an honorary doctorate from Ohio State University in 1944, declared with full sincerity:

> The research work of Mr. Midgley has received wide recognition, as is evidenced by the great number of distinctions which have come to him from those groups best qualified to evaluate his contributions to human knowledge. Through experience, the layman will also testify his indebtedness to one who has contributed so greatly to more pleasant and efficient living. He has made science a liberator, and we rejoice with him in the satisfactions that must be his in seeing the fruits of his labor. Posterity will acknowledge their permanent value.[8]

In 1947, three years after Midgley's death, his former boss Charles Kettering made a similar point during an address to the National Academy of Sciences. Kettering recalled the words of the minister at Midgley's funeral: 'We brought nothing into this world, and it is certain we can carry nothing out.' 'It struck me then,' said Kettering, 'that in Midgley's case it would have seemed so appropriate to have added this: "but we can leave a lot behind for the good of the world".'

Midgley, poor unlucky Midgley, would certainly leave a striking legacy. In his cheerful, pleasant, untiring efforts to improve the world around him, he would be inadvertently responsible for more damage to Earth's atmosphere than any other single organism that has ever lived.[9]

Midgley himself never lived to see how much trouble he had accidentally caused. In autumn 1940, he suffered an acute attack of polio, which left him paralysed in both legs. As soon as the worst was over, Midgley calculated the statistical probability that a man of fifty-one would catch the disease, and concluded that it was 'substantially equal to the chances of drawing a certain individual card from a stack of playing cards as high as the Empire State Building'. It was, he added, 'my tough luck to draw it'.[10]

Still, he continued to direct research work from his home, giving speeches by telephone and even devising a harness and pulley to lift him out of bed. But on the morning of 2 November 1944, aged fifty-five, Thomas Midgley somehow became caught up in the pulley's ropes. He was strangled to death by his own invention.

The first hints that something might be awry with Midgley's miraculous refrigerants came from a gentle, quietly spoken man with soft curls, bright eyes and a wistful smile. In spirit and interests, Jim Lovelock is far more like the natural philosophers of old than the specialised researchers of today. He is famously independent, operating – most unusually for a modern professional scientist – from a laboratory in his own backyard. He is also impishly contrary. A colleague once described him as 'the most creatively mischievous mind I've ever encountered'.[11]

Lovelock had never wanted to be beholden to a university or institution. Still, he tried the 'normal' route at least for a while. Having trained as a chemist in the 1930s, he started working for the Medical Research Council in London. But he was increasingly uncomfortable in his conventional – he would say hidebound – surroundings. While his colleagues all wore the normal white lab coats, he insisted on a surgeon's outfit, to avoid being 'uniform'. By 1959, with his fortieth birthday approaching, Lovelock had had enough. 'Every day I would go to the institute, do my research, and come home again. I felt like the man in the limerick:

> There was a young man who said, 'Damn,
> It appears to me that I am,
> A being who moves
> In predestinate grooves;
> Not a car, not a bus, but a tram.'[12]

The thought of these tramways conducting him all the way to the grave made Lovelock feel queasy. He told his boss he was quitting and fled – first to Houston, Texas, to work at the university for a while and save up some money from what proved to be a large American salary, and then back to England to set up his own laboratory in a tiny village in Wiltshire, southern England.

Lovelock quickly realised there were some practical barriers to being an independent scientist. For one thing, it was more difficult to get academic journals to take you seriously if your official address was a remote, thatched cottage instead of a prestigious institution. Trying to get hold of laboratory supplies was even worse. Even in those less terrorist-sensitive times, if you wrote from a residential address trying to order a few kilograms of potassium cyanide, say, or a piece of some radioactive substance, you were more likely to get a visit from the police than from the delivery van. To get around this problem, Lovelock decided to start up a company, which he called Brazzos. His reason for this name was typically practical. The company was named after the River Brazos, near Houston. The misspelling was because it cost twenty-five pounds a shot to compare a proposed company name with those already taken. After a couple of false tries, Lovelock picked a name he was sure nobody would have used.

Under the umbrella of Brazzos, Lovelock quickly picked up consultancy contracts from various large companies who already knew of his work for the Medical Research Council. Now he was free to pursue his inventive, sometimes outlandish, scientific ideas. Probably the most famous of these is his suggestion that life on Earth regulates its own environment to prevent the planet from getting too hot or cold, or otherwise too inimical. He came across this notion while working on experiments to detect life on Mars at NASA's Jet Propulsion Lab in 1965. While thinking about the atmosphere on Mars and our other nearest neighbour, Venus, Lovelock was struck by how different they both are from Earth. Mars is frigid, Venus searingly hot, and yet both have atmospheres whose chemistry is settled and fits sensibly with the equations. Earth, on the other hand, does not. Its atmosphere is full of ultra-reactive oxygen, for example, which by chemical rights shouldn't be there at all.

The oxygen came from life. It is there, Lovelock realised, only because living things had dramatically altered their environment to suit themselves. After that, he began finding other examples in which life had shaped the planet and been shaped in its turn. Wherever he looked, Lovelock discovered an intimate interaction between life, air and rock. It was almost, he thought, as if the planet itself were alive.

With typical romanticism, he named his theory Gaia after the Greek goddess of the Earth. (His neighbour, novelist William Golding of *Lord of the Flies* fame, suggested the name, prompting Lovelock to say that 'few scientists have had their theories named by so competent a wordsmith'.)[13] Broadly speaking, the theory was right – there are many ways that living things have adapted the planet for their own uses. However, the name Gaia, coupled with the holistic 'hippie' feel of Lovelock's theory, made many of his peers look at him askance. Though his work was both careful and sound, and published in the world's most eminent journals, some scientists still didn't trust it. Lovelock didn't particularly mind. When one eminent scientist described him as a 'holy fool'. he was even proud, though with his typical self-deprecating humour he did wonder whether the scientist had meant 'wholly'.[14]

Jim Lovelock entered the ozone story in the mid-1960s, when he became curious about the summer haze that he sometimes noticed spoiling the view from his country retreat. He couldn't remember seeing anything like that during his boyhood, and went off to visit some friends in the Meteorological Office to see whether they could explain it. Lovelock was thoroughly entertained that the UK Meteorological Office was officially part of the Ministry of Defence. He said, 'We English have always been paranoid about the weather but this seemed too much. Did we now see it as a national resource and treasure that needed the army to protect it?' And he reacted just as merrily to the news that the US Weather Bureau, by contrast, was in the Department of Commerce: 'Perhaps they thought their weather was good enough to sell.'[15]

Nobody at the Meteorological Office seemed to know what could be causing the haze, whether it was natural or had come from human hands. Then Lovelock had an idea. He knew all about Midgley's CFCs, which were now ubiquitous in English spray-cans and refrigerators. They were inert, perfectly safe, and yet could perhaps be used as a 'marker' for other, more unpleasant forms of industrial pollution. If there were higher levels of CFCs on the hazy days, that would suggest the haze was man-made.

Lovelock decided to check, and he had just the instrument for the job; like Midgley, he was a natural inventor. He made his first device – a wind gauge that he could hold out the windows of trains to measure their speed – at the age of ten, and he'd been inventing ever since. Lovelock made a good enough living from his creations to fund much of his science. But one machine was more important than most, and plays a vital role in the ozone story. Lovelock had invented a device that could detect tiny traces of many different chemicals, including CFCs. This is what he planned to use to try and determine the origin of the haze.

Back in his country retreat, Lovelock set about measuring the levels of haze, and of CFCs, on different summer days. Later that year he repeated the experiment on the west coast of Ireland. Whenever there was more haze, there were more CFCs in the air. Just as he had suspected, the haze must have come from industrial sources.

Lovelock published his results, and might have been satisfied with that. But something about the CFCs tugged at him. If they could make it to his remote village of Bowerchalke, where else could they go? They were so inert, so 'safe', that nothing could destroy them. Perhaps they were gradually accumulating everywhere in the atmosphere. Perhaps they could even be used as tracers throughout the world, tiny inert markers that would show where the harmful pollution was also going.

One way to test this would be to use his instrument to measure CFCs in the ocean, going from the relatively polluted northern hemisphere to the less polluted southern one. Direct pollution is much greater in the northern hemisphere, where there is much more land and also much more industry. So, Lovelock persuaded the national funding agency, the Natural Environment Research Council (NERC), to give him a berth on the research ship *Shackleton*, and in November 1971 he set sail.

On his first day of measurements, Lovelock hit a snag. He quickly discovered that the 'official' water samples provided for him would be useless. The problem lay not with his equipment, but with the ship itself. Because the *Shackleton* was a research vessel, seawater was automatically pumped in from the bows so that scientists could have a continuous sample for their measurements. For normal measurements that would be fine. But Lovelock's measurements were far more delicate than the ship was used to. He needed to detect his CFCs at such faint traces that even the 'clean' pipes through which the water passed were too contaminated for his purposes. He had to find another way of getting clean samples from the ocean surface.

Lovelock's first attempt at this was almost his last. He had decided on the simple stratagem of tying a bucket to a rope and dropping it over the side. However, the ship was steaming along at 14 knots, and the bucket dragged so strongly in the water that it nearly pulled him in. Lovelock was irritated with himself, and said, 'I should have calculated that a bucket dropped into water flowing past at 14 mph exerts a pull of over 100 pounds.'[16] Chastened, he asked the ship's technician for a smaller collecting bottle. But the only available ones were glass beakers from the laboratory, which would be far too fragile. It seemed it would be necessary to improvise.

Lovelock went along to the galley to see what they could offer him. Saucepans would be too difficult to manoeuvre on the end of his rope. But an old aluminium teapot, now retired from active duty, would be just the thing. From then onwards Lovelock cheerfully used this teapot to scoop up his daily water samples, to the mingled alarm and scorn of many of the ship's other scientists.

The ship's crew were more tolerant of their practical but eccentric guest, and seem to have taken his welfare very much to heart. Once, when he was collecting teapot samples in the midst of a storm, Lovelock noticed the bosun standing surreptitiously behind, ready to grab him should a wave look likely to wash him overboard.

As the ship crossed from the northern to the southern hemisphere, Lovelock noticed the difference. The air felt suddenly fresh and clean and much less hazy, and his CFC readings dropped as well. In the north they had been present at seventy parts per trillion, but the southern readings were slightly under half that. Still, the measurements proved what Lovelock had suspected: CFCs were gradually showing up everywhere.

For a research trip costing a total of only a few hundred pounds, Lovelock's *Shackleton* voyage would prove momentous. He published the results in *Nature*, and then added a rider that he would come to regret. The point of the paper was to show that CFCs were appearing around the globe, but he didn't want to raise alarm in the minds of people who were unthinkingly afraid of anything 'chemical'. CFCs are after all inert; breathing them in at a few parts per trillion would cause no harm to anyone. That's why Lovelock wrote the phrase that would later haunt him. 'The presence of these compounds,' he said, 'constitutes no conceivable hazard.'[17]

In the next few months, Lovelock's results wafted over the Atlantic to America, where they triggered a question in the mind of Sherwood ('Sherry') Rowland, a chemistry professor at the University of California, Irvine. Rowland realised that, even though Lovelock had found only tiny concentrations of CFCs in the atmosphere, together they added up to just about all of the CFCs being produced. That was odd, because most residents of the atmosphere last only a few weeks, before being reacted away or washed out in the rain. If Lovelock's measurements were right, it seemed that CFCs stayed around in the air for an extraordinarily long time. Rowland wasn't worried by this, just curious. He knew that nothing lasts for ever. What, he wondered, would eventually happen to the CFCs?

Rowland was busy with his regular research involving radioactivity, as well as all the efforts involved in running his department. Fortunately, he had a bright young post-doctoral student to whom he could hand over the problem. Mario Molina had been born in Mexico City, the son of an ambassador. His background coupled with his undoubted intelligence had opened many doors, and he had studied in some of Europe's most prestigious institutions. But he preferred the American graduate programme, and had recently finished his Ph.D. at Berkeley. He was looking for something to do next.

What Rowland proposed seemed like an interesting enough academic exercise: track the CFCs in the atmosphere, and work out what happens to them. The first thing Molina's calculations told him was that the lower part of the atmosphere held no fears for CFCs. They were insoluble in water, so they couldn't be rained back down to the ground, and there were no other reactions that could destroy them. Eventually, they would have to make it above the atmospheric ceiling that contains all our wind, clouds and weather, and emerge into the bright, rarefied levels of the stratosphere.

And that's where the trouble would start. As they floated upward into the ozone layer, the CFCs would encounter ultraviolet rays for the first time. Any stray ray that hadn't yet been soaked up by some kamikaze ozone molecule could smash into a CFC molecule instead. Like a miniature bolt of electricity, this would turn each CFC into a monster.

The danger comes because CFCs contain the element chlorine. When chlorine is safely locked up in its molecular cage it's fine. But the moment it is released by the action of ultraviolet rays, a chlorine atom begins its rampage. Through a complicated series of reactions, any one chlorine atom effectively rips off the extra oxygen atom from an ozone ($O_3$) molecule, leaving behind an ordinary molecule of oxygen ($O_2$). The same atom of chlorine could then repeat the process with another molecule of ozone, and these two extras would then react together. The upshot: two molecules of protective ozone turn into three molecules of useless oxygen[18] (or, in the language of chemical equations: $2O_3 \rightarrow 3O_2$).

But the really troublesome part was how effectively these chlorines would do their job. Each chlorine atom would finish the cycle of reactions in the same state as it had started, so it would be free to repeat the process again and again. An atom of chlorine let loose in the stratosphere was like a miniature Pac-Man,[19] gobbling up thousands, even tens of thousands, of ozone molecules before it finally reacted with something else and was taken out of the picture. According to Molina's calculations, a single chlorine atom could destroy, on average, 100,000 molecules of ozone.

Still, that would be dangerous only if there were enough chlorine atoms out there to make a serious difference to the ozone layer. Molina began some more calculations. He looked at the amount of CFCs now being released, calculated how long it would take these molecules to waft up to the stratosphere, and . . . Molina was aghast. In one hundred years' time, he calculated, the ozone layer would have lost a full 10 per cent of its molecules. He immediately raced off to see Rowland. They checked and rechecked the calculations, but the same answer kept reappearing. And 10 per cent was only a start. If their emissions remained unchecked, CFCs would pose a serious threat to all life on Earth. Rowland's mind was heavy when he returned home that night. 'The work is going well,' he said to his wife, 'but it looks like the end of the world.'[20]

The next few weeks saw Molina and Rowland going back over the figures again and again. Before they dared publish, they had to be absolutely sure of their findings. When they were confident their calculations were right, Rowland's wife Joan collected every aerosol can containing CFCs in the house and threw them all away.[21]

News of the work leaked out on to the scientific grapevine. Though he still hadn't met or talked to the pair, Jim Lovelock heard about their predictions and was intrigued. He thought it likely that CFCs would make it up to the stratosphere, and wondered whether they truly were split apart there, as the Molina–Rowland theory suggested. Never one to pass up an opportunity to test an interesting theory, Lovelock set off in search of a plane.

His first stop was the Meteorological Office, which had regular flights up to the stratosphere. But the bureaucracy there was awful. He would have to wait at least two years while the requisite safety checks could be made on his equipment and all the right papers could be stamped.

Lovelock was too impatient for that. He chatted instead to some friends at the Ministry of Defence. Did they, perhaps, know of any stratospheric flights coming up that might have room for one smallish passenger and his even smaller air-sampling cylinders? Certainly, came the reply. A Hercules aircraft was scheduled for a test flight up to 45,000 feet, and Lovelock was welcome to join it. At that time of year, the stratosphere started at 30,000 feet, so he would have three full miles of stratosphere in which to make his measurements. Of course, officially speaking he would not be on board, so there would be no compensation for his family if it crashed. On the other hand, there would be no charge, and – a blessed relief – no paperwork.

A few weeks later, Lovelock was on the flight deck as the Hercules took off from Lyneham airfield in Wiltshire. As the plane climbed, he sat next to the engineer and took his air samples. On the way down, the plane made some practice manoeuvres, including recovering from a stall. Lovelock asked rather nervously what would happen if the plane went into a spin. 'No worry at all,' came the pilot's confident reply. 'This aircraft would make no more than half a turn before the wings came off.' After that, says Lovelock, he kept quiet.[22]

As soon as Lovelock returned home he analysed his samples. They showed a steady level of CFCs in the lower atmosphere, and then a decline in the stratosphere, just as Molina and Rowland had predicted. It seemed their theory was right.

Molina and Rowland's findings appeared[23] in *Nature* in June 1974. And the reaction was . . . silence.

The two researchers had braced themselves for the onslaught that would surely follow the publication of their paper, but nobody seemed to have noticed their alarming news. The problem was they had been so careful not to overstate their case that they had buried the implications in diffident scientist-speak, with no alarm bells or warning sirens attached. 'It seems quite clear that the atmosphere has only a finite capacity for absorbing [chlorine] atoms produced in the stratosphere, and that important consequences may result . . . [with the] possible onset of environmental problems,' they had written in the middle of an obscure paragraph towards the end of the paper.

These 'possible environmental problems' involved the potential destruction of the layer that protects us from deadly space rays, but obviously this had not come through. Molina and Rowland decided it was time to make their message more explicit, both to the scientific community and to the world at large. In September there would be a meeting of many of the world's most prominent chemists, in Atlantic City, under the auspices of the American Chemical Society. This would be the perfect opportunity to present the work directly to their colleagues. But they decided they would also do something more radical: they would hold a press conference.

The relationship between scientists and the press is an uncomfortable one at the best of times. If scientists acquire anything approaching media gloss, they immediately attract both scorn and jealousy from many of their peers. There is even a name for this, the 'Sagan effect', after astronomer Carl Sagan, who – through his television programmes – woke many people around the world to the wonders of the cosmos yet was viewed with increasing suspicion by other astronomers. The general attitude among scientists is that, usually, it is better not to get involved with the media. If you must get involved, do not, whatever you do, take a political or social stand. Scientists are there to report their results. Let the world interpret them if it must.

This at least was the prevailing attitude of the time. But Molina and Rowland were about to break these rules. At their press conference, they carefully explained their results and the scientific significance. Their predictions had become gloomier; the new calculations suggested a 5 per cent ozone loss by 1995 and a 30 per cent loss by 2050. Then Molina and Rowland stepped over the normal scientific bounds. They called for a worldwide ban on CFCs.

Ban a product on which rested an $8 billion industry? On the basis of a few calculations? The CFC industry was horrified. And yet the moment was propitious for sounding warnings about potential environmental harm. The early 1970s marked the beginning of the environment as a political issue, a time when the green movement had just begun to stir into life. The Environmental Protection Agency had been born a few years earlier, after Rachel Carson had warned of the dangers of pesticides in her seminal book *Silent Spring*. Widespread excitement at the pace of technological change had given way to worry about the harm that this technology might cause. Radios around the country were broadcasting Joni Mitchell, singing her environmental fable: 'Don't it always seem to go that you don't know what you've got till it's gone?'[24]

On Capitol Hill, 11 December 1974, Representative Paul G. Rogers, Chair of the House Subcommittee on Public Health and the Environment, was introducing a hearing called in response to Molina and Rowland's findings. 'The entire matter rings of a science fiction tale,' he said. 'One we have all heard: how a planet, now barren, was destroyed by its very inhabitants. Had not the evidence been brought forth by such reputable men of science, it would seem like bitter, black humour – that the earth may be endangered and the villains of the situation are billions of aerosol cans.'[25]

Rowland is an impressive figure, six foot five, calm, imposing and supremely scientific. He laid out the case in clear, concise terms. But the problems were always going to lie in the uncertainties. Many, perhaps even most, scientists believed that the ozone layer would suffer as a result of CFCs. The trouble was that nobody had any idea how much.

The CFC industries, led by Dupont, were determined to exploit every possible weakness in Molina and Rowland's argument. And one of their star witnesses was none other than Jim Lovelock.

Why was Lovelock on the 'wrong' side? One reason is that he liked the CFC scientists he had met. Like Thomas Midgley, the people working for Dupont and the other CFC manufacturers weren't cartoon bad guys. Their companies all considered themselves to be highly reputable. Remember that CFCs had been invented only because Frigidaire – unprompted by government regulation – had decided to try and find a safer alternative to the refrigerants that were clearly dangerous. Dupont had even convened a conference in 1972 with other makers of CFCs to confirm that the chemicals weren't harmful. (Though unfortunately they confined themselves to direct health risks in the lower atmosphere, where at present levels CFCs will certainly do little harm.) Lovelock had warmed to the Dupont researchers who came to him for advice. 'Some might say I was a bloody fool,' he said later, 'but I think I just did what came naturally. I liked the people [in the CFC industry], they seemed to be a very honourable, decent bunch of scientists.'[26]

Besides, even though he knew they had a vested interest in preserving the CFC industry, he also believed that they were right. Lovelock genuinely thought that CFCs would not do a significant amount of harm. According to his Gaia theory, living things had filled Earth with self-healing mechanisms. Nature, in Jim Lovelock's book, was too powerful to be disturbed by a few whiffs of CFCs. Even if some extra ultraviolet light slipped through the ozone layer, Lovelock thought that life would be able to cope. He also had a natural distaste for knee-jerk reactions, and despised the weak-minded notion that anything 'chemical' was bad and anything 'natural' was good.

The hearing didn't achieve much, except to force the issue even farther into the open. All the scientists involved were now feeling the heat. Molina and Rowland were constantly under attack by the industry. Lovelock, meanwhile, became an environmental target. Spiteful newspaper reports in the United Kingdom began to say that he was 'in the pockets of the aerosol industry'. (It's ironic in view of all this that Lovelock's family was one of the first to give up aerosols using CFCs as a propellant. They had to – otherwise any spurt of hairspray or deodorant would have played havoc with his delicate measurements.)[27]

Later, Lovelock would change his views. He eventually realised that even the Gaian self-healing mechanisms could be overwhelmed, and also that the loss of ozone was more serious than he – or anyone else – had thought. (He was also unafraid to say so.) But for now, he was still reacting against what he regarded as unscientific hysteria. 'I respect Professor Rowland as a chemist,' he told a newspaper reporter shortly after the Molina–Rowland story broke, 'but I wish he wouldn't act like a missionary . . . The Americans tend to get into a wonderful state of panic over things like this.' What was really needed, according to Lovelock, was 'a bit of British caution'.

In April 1975, the US National Academy of Sciences put

together a twelve-person team of scientists to investigate the ozone issue. They were split into two groups. One was to review the science of Molina and Rowland's claims, while the job of the other was to make recommendations about what should be done. The teams began their laborious process of hearings and recommendations and arguments, and Molina and Rowland worried.

They were troubled because everything rested on their calculations. Apart from Lovelock's CFC measurements and a few balloons that had been sent up afterwards, Molina and Rowland had no direct measurements from the stratosphere. They had to imagine what might be happening there, and then set up artificial stratospheres in the lab to test their ideas. The stratosphere is a strange place; the air is thin, the temperature is warm and energetic ultraviolet rays abound, ripping apart the molecules that exist in more normal circumstances. Bizarre chemical species that wouldn't last a millisecond down near the ground are common in this maelstrom. And Rowland and Molina had to be sure they had included every one of the possibilities in their figures.

They knew, for instance, of two chemicals that could turn out to be either heroes or villains. Hydrogen chloride (HCl) and chlorine nitrate ($ClNO_3$) are 'reservoirs' for chlorine. They are extremely stable even in the stratosphere, and once a chlorine atom gets tied up in one of these two, its destructive habits are over. Molina and Rowland had included both of these chemicals in their calculations. But had they got it right? Give the reservoirs too much credit for their ability to rein chlorine atoms in and you'll seriously underestimate the eventual ozone destruction. Give them too little credit, on the other hand, and your results will seem like scaremongering. Until the first signs of ozone depletion showed up, nobody would know if Molina and Rowland had got it right.

Eventually, in September 1976, the reports were published. The first concluded that Molina and Rowland's calculations were justified. CFCs posed a threat to ozone. The second declared that, since it was not yet clear how serious the threat might be, it made sense to wait and see rather than to introduce immediate draconian regulations. Molina and Rowland later wrote that the two reports could have been shortened to one word each: 'yes' and 'but'.[28] Confusion reigned. Newspapers took whichever message they preferred. 'Scientists back new aerosol curbs to protect ozone in atmosphere,' declared the *New York Times*. 'Aerosol ban opposed by science unit,' was how the *Washington Post* put it.[29]

Still, a certain level of alarm had been raised. By 1978, America had at least banned the use of CFCs as a propellant. Canada, Norway and Sweden followed suit. But then, in spite of the continued attempts of Molina and Rowland and many of their scientific colleagues to keep the issue alive, ozone and CFCs dropped quietly off the political agenda. Carter was out and Reagan was in; the green 1970s had given way to the greedy 1980s.

The problem was that even Molina and Rowland now thought it could be decades before the first incontrovertible signs of ozone loss appeared. Until then, CFCs would gradually, imperceptibly nibble away at the ozone layer, and by the time we had concrete proof of a serious effect it would be too late. In summer 1984, Rowland gave a dispirited interview to *The New Yorker*:

> From what I've seen over the past 10 years, nothing will be done about this problem until there is further evidence that a significant loss of ozone has occurred. Unfortunately, this means that if there is a disaster in the making in the stratosphere, we are probably not going to avoid it.[30]

Rowland was right that nothing much would happen without new evidence of ozone loss. But he had no idea how quickly and dramatically that would come. For in the autumn of that same year, a scientist who had spent several years adopting a bit of 'British caution' decided to abandon his reticence and trumpet his findings to the world.

Antarctica in the 1950s was a rugged, macho place, and few stations were more rugged and macho than the British Antarctic Survey's remotest outpost at Halley Bay, which floats on a shelf of ice about 1,500 kilometres from the South Pole. The temperature there never rose above freezing and in the winters plunged to 50 degrees below. An even worse problem was the wind, which howled over the flat ice-shelf, whisking away any vestiges of bodily warmth as it whipped up snow into blizzard after blizzard and buried those first tough little wooden huts up to their necks.[31]

The old traditions prevailed at Halley long after they had been overtaken in the rest of the world. There were men with beards, with their arcane Antarctic-only slang and coarse humour. There were dogs to pull sledges, and there was the flat white emptiness from horizon to horizon. No room there for softness, or comforts, or women.[32]

Joe Farman, a quiet, pipe-smoking Briton of the old school, had been masterminding research at this bleak outpost since 1957. Every year, during the southern spring and summer, scientists from the British Antarctic Survey had trekked down to Halley to measure the amount of ozone overhead.

Why ozone? And why there? At first, it was an attempt to use movements of ozone to map upper atmospheric currents. Later, it was more out of habit. In the 1970s Molina and Rowland's findings about CFCs had given some extra impetus to this research, but Farman probably would have done it anyway. Long records of interesting atmospheric constituents usually prove useful in the end, for one reason or another. Compiling the records is often thankless, but after all, you never know. Farman didn't receive much money for his project, but it didn't cost that much to do and there were always plenty of volunteers to make the measurements.

Early in 1984, Farman received a visit from his boss at the funding agency, who asked him, yet again, why he was persisting with such an obscure record. 'There is a big CFC industry,' Farman replied. 'And people are writing that ozone will change. And the only way you can tell if ozone has changed is to sit and keep measuring it.' His boss's response: 'You're making these measurements for posterity. Well tell me, what's posterity done for you?'[33]

It was slightly disingenuous of Joe Farman to say this about CFCs, because he was already nursing a secret. He had been nursing it for three years, but later that same year he decided to divulge it. Something had shown up in this long, repetitive series of measurements that at first Farman didn't quite believe. Ozone always changed a bit at Halley, between the dark winter months and the return of the sun. But in 1977 something different had begun to happen. In October each year, at the onset of spring, the ozone had begun to plummet. Each year the drop was a little worse. In 1983, where Farman would normally expect about three hundred units of ozone, he was seeing less than two hundred.

At first, Farman and his two colleagues kept mum. Above all, they didn't want to look foolish. A NASA satellite had been measuring ozone over the whole of Antarctica for the past five years and had noticed nothing amiss. Perhaps there was something funny about the instruments Farman and his group were using. Perhaps there was something funny about Halley itself. So in the season of 1983–84, Farman sent a new instrument down to Halley. He also took a look at the record from another British station, Argentine Islands, which was more than a thousand miles farther north. Both stations confirmed what Halley had already shown. Now 40 per cent of the ozone was disappearing each austral spring. There was a hole in the sky.

When he saw this, Farman threw his British caution to the winds. The paper that he and his two colleagues wrote landed in the offices of *Nature* on Christmas Eve and was published in

May 1985.[34] Molina and Rowland's paper had had little immediate effect, but Farman's caused an uproar. Among the most astonished was Donald Heath's research group at the NASA Goddard Space Flight Center, whose job was to co-ordinate the ozone measurements made by NASA's Nimbus-7 satellite. They had no hole in their data. What was Farman's group talking about?

Hastily, Heath's group pulled their data out to check it again. They were mortified. The data recovery programme had been designed to throw out spurious numbers before the researchers even saw the results; that way they wouldn't have to be bothered by irritating measuring glitches. Any measurement of ozone values that fell below 180 units was obviously ridiculous and had simply been ditched. The satellite had seen Farman's ozone hole all right, but thanks to their overeager programme the researchers themselves hadn't seen a thing. Now, using the correct data from 1979 to 1983, they watched a hole the size of the continental United States gradually appear over Antarctica. In some cases, the ozone dropped to less than 150 units.

Heath's group had learned an important lesson about Earth's atmosphere. Even if you are sure you understand the way our ocean of air works, it is still always wise to expect the unexpected.[35]

Meanwhile, the rest of the ozone community was in disarray. Even Molina and Rowland's worst scenarios hadn't predicted something as extreme as this, and so soon. There was no sign of a hole like this anywhere else on Earth, so it must have something to do with the extreme Antarctic conditions. But what?

In research labs and coffee rooms at universities throughout the world, attention began to focus on the first, most obvious characteristic of the Antarctic stratosphere: it is the most isolated air on Earth. Every winter, winds whip up around the edge of the entire ice-covered continent until they form a giant vortex whose walls separate the air from warmer breezes farther north. Trapped inside this vast whirlwind, Antarctic air grows steadily colder, and colder. And then, a new kind of cloud appears in the Antarctic skies.

Normal clouds are formed of liquid water drops, and they can occur at almost all levels in the troposphere – the lower part of the atmosphere – which is the place where we live and experience our wind and weather. As you progress upwards through this layer of air, the temperature drops steadily until, at the top of the troposphere, it reaches a minimum. Immediately above this point, the stratosphere begins. Now there are ozone molecules to catch sunlight and warm the air, and the temperature starts to rise. The cold point between these two layers traps any water vapour by turning it into clouds and sending the rain falling back down to Earth. It is a watertight barrier, stretching around the world like a giant tarpaulin, keeping the lower atmosphere wet and the upper atmosphere bone-dry. That's why the stratosphere almost never has clouds.

But the stratosphere still contains just a little water that has leaked through from below, and – if the temperatures are low enough – this can freeze solid into tiny flecks of ice. That is what happens in the Antarctic stratosphere, in winter.

They are beautiful, these clouds, iridescent like the inside of an abalone shell with hot pinks and purples and shimmering blues, colours that don't belong in the sky. In spring, when the sun has returned after the long polar night, they materialise at sunrise or sunset seemingly out of thin air. In fact, the clouds are there all the time, but it's only as the sun tips over the horizon that it picks them out, shining on them like a spotlight with its final steep rays. And then, suddenly, it's as if the sky is filled with glimmering peacock feathers. The early explorers made exquisite watercolours of the effect. They had no idea how dangerous it would turn out to be.

Many different researchers began to suggest that these high stratospheric clouds might explain the Antarctic ozone hole. Among them was a thirty-year-old theoretician named Susan Solomon, who was working at the National Oceanic and

Atmospheric Administration in Boulder, Colorado. Though she was young, Solomon was very talented. She had been one of the original reviewers of Joe Farman's paper, and when she read it she was immediately horrified. Ever since, the data had nagged at her.

Hunched over her computer, Solomon had tried model after model, accounting for every reaction she could think of in the Antarctic stratosphere. None of them made a hole of this magnitude. Then she began thinking about the clouds. What if they made the difference? Perhaps they somehow 'primed' the Antarctic atmosphere during the winter, so that when sunlight returned in spring the destruction could begin.

Cloud surfaces can make a big difference to chemical reactions, especially in somewhere as thin as the stratosphere. For any chemistry to happen in the air, two atoms or molecules have to meet. But up where the air is rarefied, such encounters don't happen very often. What's more, the two, or more, participants in the reaction need to be suitably energised. If they're overly lethargic, nothing much will come of their meeting.

But if any of these chemicals can land on the surface of a cloud, they immediately have many more options. The cloud can act as an introduction agency, both bringing species together and giving them an energy boost to get them going. That's what Solomon found when she began to put clouds into her model.

The key seemed to be in those unreactive 'reservoir' species that could bind up chlorine atoms and keep them out of trouble. Throughout the long winter nights, these molecules – chlorine nitrate and hydrochloric acid – could be landing on the surface of clouds and reacting. Work through the sequence of reactions and you end up with chlorine gas, $Cl_2$. This molecule detaches itself from the clouds and awaits the return of sunlight. The first ultraviolet rays that appear with the rising sun split the chlorine gas into its individual, deadly, atoms, which rip ravenously through the ozone layer. Suddenly a hole made perfect sense. The only wonder was that it wasn't even worse.

Still, this was only a theory. Around the United States, many other research groups had come to similar conclusions, but they all knew that nobody could be sure until they had more data. Someone would have to go south to Antarctica, to measure the reactions as they happened at the end of the twenty-four-hour darkness of winter and on into the first few weeks of spring.

Solomon was as surprised as anyone to find herself volunteering for the job. She was a theoretician. Her work involved sitting at a computer, not going out into the field. For her first-ever venture into experimental science, she would lead a team of twelve people to the coldest, most hostile place on Earth, in the winter.

She still can't explain why she wanted to go. Perhaps because, in spite of the sedentary nature of her chosen profession, she was captivated by the might of the atmosphere. She loved storms, thunder, lightning – anything that reminded her how powerful nature is, and how puny humans are by comparison.[36]

For whatever reason, a few months later Solomon found herself at the airport in Christchurch, New Zealand, sweltering in her red fur-trimmed parka, clutching her standard-issue canvas bag full of safety gear and survival clothing. Ahead was a hazardous eight-hour flight in a Hercules military transport plane. The pilot arrived for the briefing and regarded the twelve men and one young woman in front of him. 'Who's in charge here?' he asked. Solomon raised her hand. The pilot regarded her with some astonishment and then managed to stammer out, 'Good for you.'

Solomon loved Antarctica from the moment she stepped off the plane. She loved the emptiness, the ferocity, the unforgiving wildness. It wasn't beautiful in the usual, picture-postcard sense. Its beauty was fierce, and Solomon revelled in it.

She had arrived at McMurdo, the main American base, which is the unofficial 'capital' of the continent. It was August 1986, the tail end of the austral winter. The influx of summer visitors would not begin for another month or so, when the weather warmed and daylight began to stretch towards twenty-four hours. The only occupants of the base were the people who had been isolated there for the whole winter, who had bonded tightly with each other into established cliques and enmities and tended to regard with suspicion newcomers who would unwittingly take the wrong chair at dinner, or hang their coat on 'somebody else's' peg.

Solomon's task was to set up an instrument on the roof of one of the buildings. The idea was to use incoming moonlight to pick out and measure the chemicals floating in the intervening stratosphere. The measuring instrument would be inside the building, but up on the roof would be mirrors that would turn to guide the moonlight down into a channel.

The team had only four months' notice of their journey. They hadn't had time to construct a tracking system to turn the mirrors as the moon journeyed across the sky. Instead, someone had to be on hand, braving temperatures of –40 degrees and the occasional fierce wind that could blow up almost out of nowhere.

One night, Solomon was up on the roof taking her turn when the weather turned cloudy. Without moonlight the instrument was useless. Solomon decided to leave the mirrors and climb back down into the lab for a nap. Perhaps when she returned, the moon would be back. In the laboratory, Solomon curled up in her sleeping bag and fell asleep. She woke to a blizzard, the worst kind of white-out, winds screaming past the building with a ferociousness rarely seen outside the ice. Solomon was aghast. Her mirrors were still on the roof. If they became damaged, the project was over.

Without stopping to think, she climbed back up the wooden ladder on to the roof, bracing her face against the shards of snow that blasted like sand. She flung herself, spread-eagled, on to the roof surface and began to edge her way towards the mirrors, the gusts tugging at her, urging her to fall. But she held on to the mirrors, and the ladder, and managed to scramble back inside.

It was worth it, she says. It was all worth it. Because Solomon's research, and the studies that followed, showed beyond doubt that high stratospheric ice clouds were indeed doing the damage. Each winter, they took the chlorine reservoirs and activated them, priming the Antarctic air like a grenade. True, this is a problem that is unique to the unoccupied continent of Antarctica; even the Arctic doesn't get cold enough to form stratospheric clouds for long, and there has never been an ozone hole in northern parts. But though you could argue that the Antarctic problem would affect only a few penguins and scientists, the striking image of deadly rays flooding through a hole in the sky turned the ozone tide.

On 16 September 1987, under the auspices of the United Nations Environment Program, twenty-one nations and the European Community signed the famous Montreal Protocol, the first international agreement ever made to restrict the emissions of an environmentally hazardous material. There would be a 50 per cent reduction in CFC production by the end of the century.

In March 1988, new analyses of ozone measurements over the United States, Canada, Japan and northern Europe revealed that, though not as severe as the Antarctic loss, the air was thinning in the north as well. Two weeks later Dupont, the world's largest CFC manufacturer, announced it would cease production.[37]

As the ozone hole continued to deepen and yet more scientific evidence flooded in to link ozone loss to Midgley's CFCs, the targets grew more stringent. In 1990 an amendment signed in London required a complete ban by the end of the century. Two years later in Copenhagen, the rules changed yet again. Now there would be a ban on CFC production and use by 1996.

In 1995, Molina and Rowland were awarded the Nobel prize for their work identifying the dangers of CFCs.[38] The other researchers involved received their share of awards and accolades. Solomon even had a glacier named after her in Antarctica. When she first heard about this by fax she thought it was a joke, and that the glacier had in fact been named after some Antarctic explorer who shared her surname. It was only after she left the fax in her in-tray for a week that she read the small print and realised it was true. She now describes this as her 'favourite honour'. And all had a personal prize that comes rarely in scientific research: the knowledge that their work has helped save the world.

Midgley's monsters are very long-lived. They will remain in the atmosphere throughout the twenty-first century; you will inhale some of them in every breath you ever take. The ozone hole will continue to appear every spring over Antarctica, too, and will probably get worse before it's better. In the end, though, some time towards the end of the century, the hole itself will heal, and our protective shield will be back in place.

There is one last sting in this tale. Many people confuse the twin environmental bugbears of the ozone hole and global warming, though they are actually independent problems, each with its own separate cause. And yet there is a sinister connection between them. Global warming makes the tarpaulin water-barrier between the troposphere and stratosphere just a little more leaky, so that a warmer world will contain a stratosphere that's slightly more damp. Also, warming in the troposphere means that the stratosphere gets cooler. Put these two together and the conditions become even more favourable to make more stratospheric clouds, not just in Antarctica, but also in the north.

Until now the Arctic has been protected from an ozone hole. The surrounding mountainous landmasses disrupt the air flow, which stops a true vortex from forming, so it's never quite cold enough to make stratospheric clouds for long. But global warming could yet change that. For three months from the end of November 2004, there were more stratospheric clouds over the Arctic than have ever been seen, and they persisted for longer than usual. And in spring 2005, some 50 per cent of the ozone layer disappeared overhead. Though this wasn't quite a hole on the Antarctic scale, it has much more chance of reaching inhabited regions. Unlike the tightly isolated Antarctic atmosphere, the northern vortex tends to slew around like a wobbling top; in the same year, it drifted down over northern Europe as far south as Italy.

Perhaps we all need to bear in mind the words of Jim Lovelock, who now fully appreciates the dangers of CFCs. In 1999, approaching his eightieth year, he wrote this:

> Our planet is one of exquisite beauty: it is made of the breath, the blood and the bones of our ancestors. We need to recall our ancient sense of the Earth as an organism and revere it again. Gaia has been the guardian of life for all its existence, and we reject her care at our peril.[39]

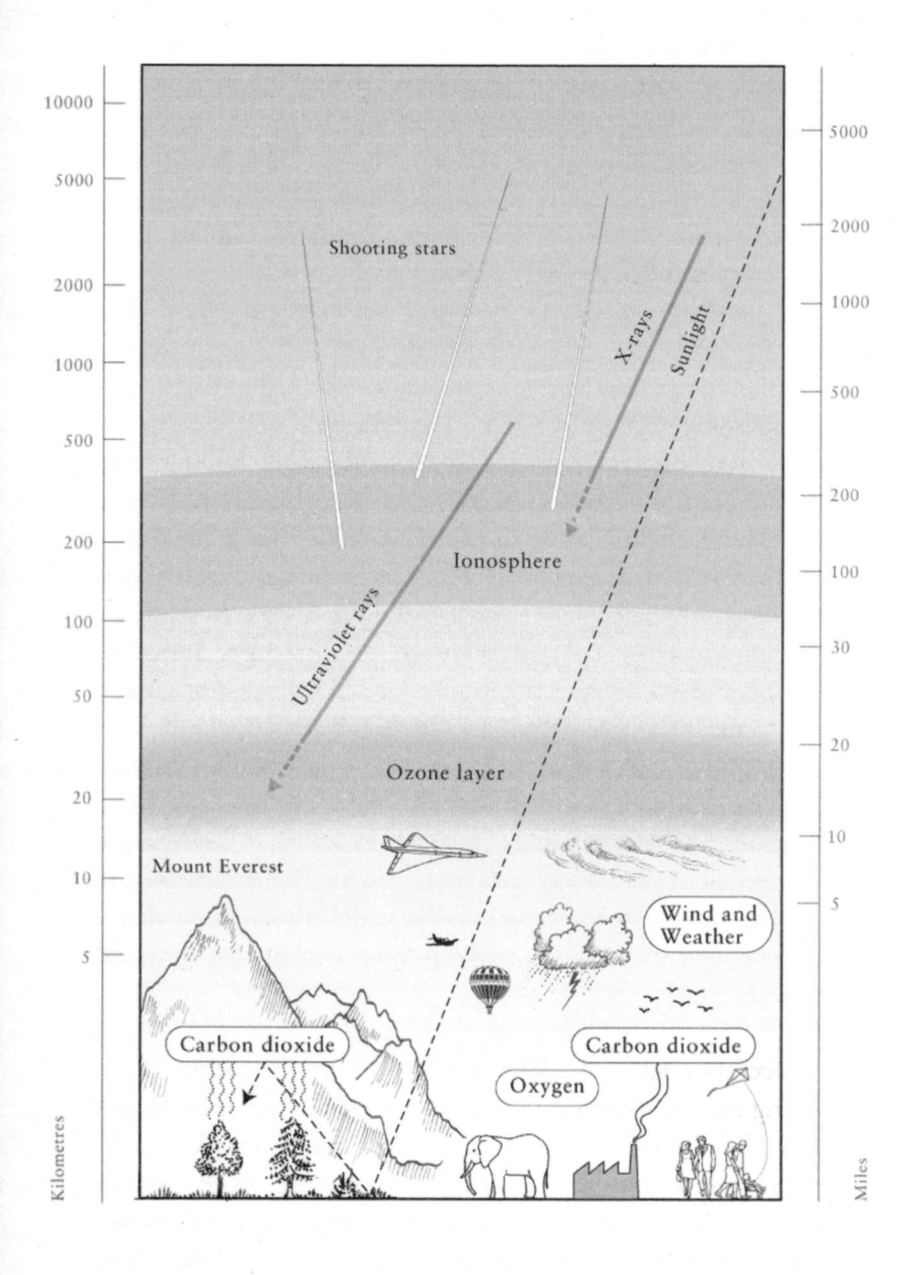

10000
5000
2000
1000
500
200
100
50
20
10
5
Kilometres
5000
2000
1000
500
200
100
30
20
10
5
Miles
Shooting stars
X-rays
Sunlight
Ionosphere
Ultraviolet rays
Ozone layer
Mount Everest
Wind and Weather
Carbon dioxide
Carbon dioxide
Oxygen

*Chapter 6*

# MIRROR IN THE SKY

Starting some 100 kilometres above Earth's surface, the air crackles with current. This is a mysterious region of our atmospheric ocean. It is the home of shooting stars and strange dancing jets of light – some long, thin and blue, which are drawn all the way up from the tops of thunderclouds deep below them; others, gigantic blobs of red with flailing tentacles. Researchers have only recently spotted these weird ultra-high forms of lightning and have given them appropriately whimsical names: elves, sprites and goblins.

They provide the backdrop to the most important function of this high electrical layer: it is the big brother of the ozone layer, soaking up rays from space so deadly that without it Earth would be lifeless. The first indication that this high electrical region existed at all came from someone who hadn't the slightest clue it was there, but was still hoping with all his heart that it would help him.

**12 December 1901, 12:30 P.M.**

A young man was sitting at a desk in a small building perched on Signal Hill, Newfoundland. Though the room was dusty, the table in front of him bore the highest technology of the time: a curious jumble of leather boxes and shiny gold wires, and a small bronze device that the man had pressed to his ear. He knew, or at least he hoped, that 2,200 miles away in Poldu, Cornwall, a team of workers were cranking up their aerials to broadcast a message to him. But all he could hear was crackles.

Guglielmo Marconi was an unusual mix. His Irish mother, daughter of the wealthy Jameson whiskey family, had run away from home to marry her Italian beau, against whom her parents were entirely opposed. Their antipathy was understandable. Annie was only twenty-one, and Giuseppe Marconi was thirty-eight. Worse still, he was a widower, who already had a son of his own. He was foreign, too, living in some far-off mountainous region that had little to do with the vibrant society circles in which Annie's family habitually moved.

But Annie had been determined. And although Giuseppe grew a little more distant over the years, she never regretted her elopement. Her first son, Alfonso, was born a year after the marriage; the second, Guglielmo, came a full nine years later, in April 1874. Perhaps because of her husband's increasing remoteness, Guglielmo had all of Annie's heart. According to family legend, among the servants crowding into Annie's room to view the new baby was an old gardener who blurted out, 'What big ears he has'. Annie is said to have retorted, 'He will be able to hear the still, small voice of the air.'[1]

Marconi's father was nearly fifty when he was born, and had little patience for mewling infants. Giuseppe's other two sons caused him no trouble. They were quiet and obedient. They submitted to his strict authority with respect. But Guglielmo was in trouble with his father almost from the moment he could talk. At mealtimes, when children were supposed to present themselves punctually, suitably scrubbed and ready to engage in informed conversation only when called upon to do so, Guglielmo was inclined to arrive late, splattered with mud or dust, and to blurt out whatever new ideas were unaccountably running through his head.

Unlike her husband, Annie had great respect for the rights of children. 'If only grown-ups understood what harm they can do to children,' she would later tell her granddaughter, Degna.

'They think nothing of constantly interrupting their train of thought.' Annie saw that her young Guglielmo was full of ideas. She also saw that he was capable of intense, obsessive concentration, and she did her best to ensure he had the space and time to put his schemes into practice.[2]

Marconi's early life was full of these contradictions between his parents' characters and cultures: Italian or Irish, Catholic or Protestant, strict or indulgent. But among these different languages, religions and attitudes between his father's stern criticisms and his mother's warm indulgences, there was always space for Guglielmo to slip through the cracks and have his own way. His unconventional upbringing only served to accentuate the characteristics he was born with. He grew up to be reserved, yet wilful, focused, independent and never, ever daunted.

As relief from the formal lessons that he hated, Guglielmo often escaped to his father's library. His early love of Greek mythology soon gave way to the enthralling works of Benjamin Franklin and Michael Faraday's classic lectures on the new science of electricity. These readings fuelled Marconi's passion for fiddling with machines. Before he was ten, he had taken apart his cousin Daisy's sewing machine and rigged it to a roasting spit. (When Daisy cried, the repentant Guglielmo immediately put the machine back together.) At thirteen he made a secret still for distilling spirits. But it was the experiment with the dinner plates that destroyed any remaining shreds of his father's patience. Guglielmo had read all about Franklin's experiments with electricity and decided to try one of his own. For some reason, he decided to wire up a series of plates, but when he sent a jolt of electricity through them, the plates all crashed to the ground. To Guglielmo's father, this was wanton destruction of the most wasteful kind, and for no conceivable purpose. From now on, if he came across one of his son's infernal contraptions, he destroyed it. And from now on, Guglielmo coolly made sure that his inventions remained well hidden.[3]

At first the young Marconi was just playing with his inventions, but in summer 1894, while on vacation in the Italian Alps, he read something that would change his life. A German scientist named Heinrich Hertz had recently died, and Marconi stumbled across his obituary. This didn't only write about Hertz's life, it also explained some of his scientific work.

Seven years earlier, it seemed that Hertz had discovered something extraordinary: invisible electromagnetic waves. A brilliant Scottish scientist named James Clerk Maxwell had predicted that such waves existed, but until Hertz nobody had seen them in action. They had crests and troughs like normal waves, but travelled at the speed of light. In fact they were simply stretched (or squeezed) versions of light waves, with wavelengths too long (or too short) for our eyes to see.[4]

Hertz had made the waves using a loop of copper wire, its two ends separated by a gap several inches long. When he pressed a key to connect the wire loop to a source of high-voltage electricity, a large blue spark leapt across the gap. The spark itself was normal visible light; there was nothing mysterious about that. But it also set off an electrical disturbance in the surrounding air, a little like dropping a stone in a pond. 'Ripples' of alternating electric and magnetic waves spread out invisibly from Hertz's apparatus.

He knew this because he had also stationed a second loop of wire a few feet away, to act as the receiver. The gap in the receiving loop was much smaller, just a fraction of an inch. But when the first spark leapt, invisible waves did indeed shoot over to the receiver and a tiny blue spurt appeared across the gap there, to show they had arrived.

When Marconi read about this experiment, it triggered a spark in his receptive mind. Perhaps these new Hertzian waves could be used to send messages.

Thanks to the Industrial Revolution, long-distance communication was heavily in demand. At first the technology was very rudimentary. The beginning of the nineteenth century had seen a profusion of semaphore stations spring up throughout Europe. Each station had a tall post bearing a pair of adjustable arms. At the next station, an operator would peer through his telescope, mark down the message's shifting letters, and then adjust his own semaphore arms to pass the information on. Though the process was laborious, a message could pass this way from Paris to the coast in just a few minutes. Better still, in 1860, when most stations had been refitted with the new electric telegraph, cables were laid underground or carried on poles to transmit the dots and dashes of Mr Morse's code wherever they were needed.

But there the technology had stalled, because messages could go only where the cable was laid. Hence the essence of Marconi's idea: if these Hertzian waves could be made to travel far enough, they could enable communication from anywhere to anywhere, eliminating the need for cables. True, Hertz's waves had travelled only a few feet, and had been greatly enfeebled by the journey. But surely he, Guglielmo Marconi, could improve on the design. Back home he persuaded his mother to let him use two attic rooms for his laboratory, and he worked on his invention for the whole of the winter. Marconi had failed both his Navy entrance exams and his attempts to matriculate at the university, and his father had all but washed his hands of this useless son. But Marconi's genius would never lie in theory. His approach was all practice.

By spring 1895, Marconi had a working system that could send and receive dots and dashes across his attic laboratory. By summer he had moved his 'wireless' out of the house. He began sending messages hundreds of yards, down to the fields in front of the villa where his older brother Alfonso was waiting with the aerial. Alfonso would wave a handkerchief attached to a pole as soon as the message arrived, and the flutter of white cloth was appearing at greater and greater distances. Still, Marconi knew that his invention would never amount to much unless he could use it to communicate across natural obstacles like hills and mountains.

With no reason to believe that it would work other than his inner conviction, Marconi resolved to send a message to the other side of a nearby hill. For this, a handkerchief would be a useless signal. Loyal Alfonso took up a hunting rifle and headed off up the hill, followed by two assistants hauling the antenna. It was a golden September day. Marconi watched for twenty minutes until the procession mounted the brow of the hill and then disappeared from view over the horizon. He waited a few minutes more, then sent his message. In the distance, an answering shot from Alfonso's gun echoed down the valley.[5]

Marconi had no idea why or how the waves had passed over the hill. He thought at first they had somehow travelled through it. But he had proved that his wireless waves really could go far. One consequence of this experiment was that, at last, Marconi's father was impressed. All his criticisms evaporated. Now his son's invention was something he could understand, a business proposition. He gave him the money he needed to develop it further, and even sought support for it from the Italian government.

To everyone's disappointment, the government declined. So, Marconi travelled with his mother to London, where her family connections would get him a hearing, and where he was better received. In spite of Marconi's funny foreign name, London businessmen were reassured to find that he carried himself in the manner of a perfect English gentleman. He spoke slowly and carefully, but – thanks to his mother's lessons – his English was flawless, without the slightest trace of an Italian, or even an Irish, accent. He had his mother's steady blue eyes, her light hair and fair complexion. His air was surprisingly confident for one so young, yet he was scrupulously neat and neither flamboyant nor showy, either of which attributes would surely have scared away the sober London investors.

(While London decided that Marconi was comfortingly English, America would later find him pleasantly continental. 'When you meet Marconi,' one US reporter later observed, 'you're bound to notice that he's a "for'ner". The information is written all over him. His suit of clothes is English. In stature he is French. His boot heels are Spanish military. His hair and moustache are German. His mother is Irish. His father is Italian. And altogether there's little doubt that Marconi is thoroughly a cosmopolitan.'[6] He was no smooth sophisticate, though. Another American newspaper reported: 'He is no bigger than a Frenchman and not older than a quarter century. He is a mere boy, with a boy's happy temperament and enthusiasm, and a man's nervous view of his life's work. His manner is a little nervous and his eyes a bit dreamy. He acts with the modesty of a man who merely shrugs his shoulders when accused of discovering a new continent. He looks the student all over and possesses the peculiar semi-abstracted air that characterizes men who devote their days to study and scientific experiment.')[7]

London businessmen were impressed by Marconi's invention as well as his manner. Unlike the Italian government, they were quick to see the possibilities, and very ready to put up the money to exploit this new 'wireless' technology. On 20 July 1897, Marconi became majority shareholder in a new company, the Wireless Telegraph & Signal Co. Ltd (later renamed Marconi's Wireless Telegraph Co. Ltd). As well as 60 per cent of the shares, Marconi received fifteen thousand pounds in cash. He was twenty-three years old, rich, and beginning to become famous.

The following summer, Marconi went to Poole, on the south coast of England, to participate in the first journalistic use of wireless telegraphy. The Dublin *Daily Express* had sent their sailing correspondent to follow the regatta of the Royal St George Yacht Club. He watched the race from the bridge of a tugboat, and then passed his reports on slips of paper to where Marconi sat at his wireless transmitter.

The reporter found Marconi refreshingly candid about his inability to explain the mysterious behaviour of his invention. He wrote that when Marconi was working on the transmitter, 'his face shows a suppressed enthusiasm which is a delightful revelation of character. A youth of twenty-three who can, very literally, evoke spirits from the vasty [*sic*] deep and dispatch them on the wings of the wind must naturally feel that he had done something very like picking the lock of Nature's laboratory. Signor Marconi listens to the crack-crack of his instrument with some such wondering interest as Aladdin must have displayed on first hearing the voice of the Genius who had been called up by the friction of his lamp.'[8] Another reporter on board confessed that it was almost irresistible to play with the wireless. 'No sooner were we alive to the extraordinary fact that it was possible, without connecting wires, to communicate with a station which was miles away and quite invisible to us, than we began to send silly messages, such as to request the man in charge of the Kingston station to be sure to keep sober and not to take too many "whisky-and-sodas".'[9]

The race went on for two days. Sometimes the fog was so dense that the ships were invisible from shore, and the only news came from the steady trail of Morse code from Marconi's wireless transmitter. The newspapers were full of this amazing new demonstration of the nautical power of wireless. Queen Victoria herself heard about their reports. The queen, now nearly eighty, was staying at Osborne House on the Isle of Wight, and she asked Marconi to set up a receiving station on the royal yacht, moored offshore, so she could communicate by wireless with her son, Edward, Prince of Wales, who was living on board. More than 115 messages of great significance passed among the royal household in this way. The queen was able to ask her son if he had slept well, and members of her entourage sent invisible wireless messages through the ether to invite members of the prince's entourage for tea. Marconi was awarded a 'handsome scarf pin' by the prince in recognition of this particular service to the nation. And the press was once again enchanted.

All the while, Marconi was finding new ways to improve the power of his transmitters and the sensitivities of his receiver.

From his permanent stations on the south coast, he was sending his messages farther and farther; nothing seemed to stop them. Even Earth's curvature didn't seem to stand in their way. This was particularly surprising since Hertzian waves were supposed to travel in a straight line. Any that skimmed the horizon should simply shoot out into space.

And yet Marconi's messages didn't seem to care. Though the lighthouse at the Isle of Wight was some 100 feet above sea level, the curving Earth meant that its tip was barely visible from the mainland at Poole. But wireless waves skipped effortlessly over this apparent barrier. Then Marconi sent messages from ship to a shore that was fully 25 miles away, over an intervening 'hill' of water that was effectively 500 feet high.

This awakened an extraordinary idea in Marconi's mind. Could wireless solve the problem of communicating between ships? This was the dawn of the twentieth century. Yet while telegraph wires had connected entire continents, ships were still forced to rely on signalling techniques that were embarrassingly and hopelessly ancient. Flags, flickering lamps and semaphore were the only voices a ship could use, and the moment it disappeared from sight, it lost all touch with the world.

Until the twentieth century, one commentator later wrote, 'ships burned or foundered in storms with not so much as a whisper reaching land to tell their fate . . . Wireless telegraphy with its magic powers was to wrest from the sea its ancient terror of silence and to give speech to ships which had been mute since the dawn of navigation.'[10]

That was Marconi's dream. But to achieve it, he needed to prove that the mathematicians were wrong, and that wireless waves really could span mighty curving distances. Twenty-five miles wasn't a stirring enough demonstration of wireless's powers. Marconi decided to send a wireless message over a mountain of water that, because of Earth's curvature, would seem more than 150 miles high. He would prove that wireless could compete with cable, even across the three thousand miles of the Atlantic Ocean. The mathematicians of the time were still declaring this to be impossible, but Marconi adopted the same policy as with every other experiment: I believe it will work. Let's see if I'm right.

The Anglo-American Cable Company was watching Marconi's work with hostile attention. They were the owners of the fourteen cables that presently spanned the floor of the Atlantic Ocean. Each one had cost around $3 million to lay, and though the price of sending messages along their vast lines was beyond the reach of most people, the cables were all working to full capacity transmitting Morse blips between the continents. If wireless worked, however, cable would be out of business. The commercial stakes were huge. Marconi felt it imperative that he say nothing until he had proved that waves could leap the 150-mile mountain that was the curving Atlantic Ocean.

Thus, in January 1901, a London newspaper reporter asked Marconi, 'Is there any truth in reports that you are contemplating the sending of messages between this country and America?' 'Not in the least,' Marconi had replied. 'I have never suggested such an idea and though the feat may be accomplished some day, it has as yet hardly been thought of here.'

Yet that same month, Marconi took a map of America and marked out Cape Cod, where 'a man can stand and put all America behind him'.[11] This, he decreed, was to be the place where the first transatlantic wireless signal would be heard.

The British side of the transmission was to be at the foot of Cornwall, from a town called Poldhu (pronounced 'Pol-ju'). Engineers had already begun to erect the gigantic transmission aerial that Marconi believed he would need. He had designed a semicircular structure containing twenty massive wooden masts, each 200 feet high and strung around with wires. But after eleven months of hard effort, just as the structure was near completion, disaster, in the form of a gale, struck the windswept Cornish coast. The masts were tall, but not strong. They tumbled like dominoes.

A few weeks later, in October 1901, a storm tore through the receiving station at Cape Cod. One of the great pine masts punched through the roof of the transmitting hut, narrowly missing one of Marconi's chief engineers. Now, the receiving aerial lay smashed on the shore like so much driftwood.

Marconi swiftly changed his plans. He ordered a new, simpler and more robust transmission aerial to be built at Poldhu. It would have two masts, not twenty-four, to be strung at either side with fifty-five lengths of copper wire, like guy ropes on a tent. He waited just long enough to test the new aerial, having it broadcast messages to a station at Crookhaven, Ireland, which was 225 miles away. This marked a new record for wireless, but Marconi scarcely noticed.

On 26 November 1901, he boarded the liner *Sardinia* at Liverpool. He had changed his mind about the receiving station. The liner was bound not for Cape Cod, but for Newfoundland, which was the closest point on the North American continent to England. With him were two of his most trusted engineers, Percy Paget and George Kemp, each of them sporting a splendid black handlebar moustache. (Marconi's own moustache was much more discreet.) It was far too late in the stormy Atlantic season to think about erecting another set of vulnerable aerial masts, but Marconi had no intention of waiting until spring. Instead, he had decided to try a different strategy for creating an instant, 600-foot aerial. In his luggage were six kites, two balloons and a very large amount of copper wire.

After looking over several sites in St John's, Newfoundland, Marconi picked Signal Hill, a high crag overlooking the port, which protected St John's from the fierce Atlantic storms. Atop the hill was a small plateau, which would be suitable for kite-flying. There was also Cabot Memorial Tower, used for signalling ships, and a two-storey stone building that was formerly a military barracks, now used as a hospital.

Marconi was about to try to extend his distance record by a factor of ten. The weather was awful: wind and sleet slammed against the walls of the hospital building. Marconi and his team inflated a balloon with hydrogen to see if it could rise high enough, trailing its ten pounds of wire beneath it. But the winds were so strong that the balloon's heavy mooring rope snapped like a piece of thread and it disappeared out to sea.

On Friday, 12 December 1901, Marconi decided to try again, this time with a kite bearing an aerial wire 600 feet long. He cabled Poldhu with his instructions. They were to send the letter 'S', dot-dot-dot in Morse code, beginning at 11:30 Newfoundland time and continuing for the next three hours.

A little before noon, the experiment began. Paget was outside, battling with the kite's rope. Icy rain fell on his upturned face as he watched the kite struggle with the wind, now surging up to 400 feet, now tumbling again to just above the heaving Atlantic.

Inside the small dark room, Marconi's other assistant, Kemp, was sitting at the single chair. In front of him on the table were a few coils and a condenser. Around him were only packing cases. Marconi swallowed a cup of cocoa and then took his turn at the receiver. He placed a single earphone in his ear and started listening. It was now just after midday.

Marconi was fully focused on his task. One reporter said of him that no portrait could convey 'the peculiar lustre of his eyes when he is interested or excited . . . One of the first and strongest impressions that the man conveys is that of intense activity and mental absorption.'[12] He had risked more than fifty thousand pounds to prove something that had been declared impossible by many of the world's leading physicists. All he had on his side were the short hops that the waves had managed back in Europe, and his absolute confidence that he was right.

He listened intently, but the minutes passed and there was nothing but crackles. Then, at half past twelve, after more than an hour of broadcasts from the team at Poldhu, Marconi heard something. Surely, three sharp clicks had sounded in his ear. He handed the earphone to Kemp. 'Can you hear anything, Mr Kemp?' he asked, calmly. Yes, Mr Kemp could indeed hear the same three clicks. Paget was immediately called in. He heard nothing, but he was, after all, slightly deaf. And then, though Poldhu had been instructed to send the signals continuously, they suddenly stopped.

Marconi continued his vigil. At 1:10 the signals repeated, and again at 1:20. By the end of the day's broadcasting period, Marconi had heard that Morse 'S' twenty-five times.

But he still wasn't sure if that was enough. The signals had been faint and erratic, and Marconi wanted something a little more substantial. The next day, Saturday, 13 December, the team tried again. This time, however, the wind was too strong, and the kite aerial was useless. The weather was getting worse. Marconi decided that he'd done enough. He had definitely heard the signal, and now was the time to issue a formal statement.[13]

The press went wild. Under the heading 'Wireless Signals Across the Atlantic – Marconi says he has received them from England',[14] the *New York Times* wrote, 'Guglielmo Marconi announced tonight the most wonderful scientific development in modern times.'[15]

A reporter in *McClure's* magazine captured the public wonder at the announcement. 'A cable, marvelous as it is, maintains a tangible and material connection between speaker and hearer; one can grasp its meaning. But here is nothing but space, a pole with a pendant wire on one side of a broad and curving ocean, an uncertain kite struggling in the air on the other – and thought passing in between.'[16]

Even many of the usually sober scientists could hardly contain themselves. Sir Oliver Lodge, a British pioneer of electrical research, wrote: 'One feels like a boy who has been long strumming on a silent keyboard of a deserted organ, into the chest of which an unseen power begins to blow a vivifying breath. Astonished, he now finds that the touch of the finger elicits a responsive note, and he hesitates, half-delighted, half-affrighted, lest he should be deafened by the chords which it seems he can now almost summon at his will.'[17]

Stocks in cable companies plummeted. The frantic efforts of the Anglo-American Cable Company to discredit Marconi's experiment only made his results seem more reliable. None of his peers doubted him.

A few weeks later,[18] the American Institute of Electrical Engineers held a glittering dinner for Marconi in the Astor Gallery of the Waldorf-Astoria in New York. Behind Marconi's table was a black tablet studded with electric lights spelling out his name. At the eastern end of the gallery, the word 'POLDHU' was similarly picked out in lights, and at the western end, 'ST JOHN'S.' All three tablets were linked by a cable bearing electric lamps, bunched together in threes to form the letter 'S' in Morse code. The toastmaster read out a letter from the great inventor Thomas Alva Edison: 'I am sorry not to be present to pay my respects to Marconi. I would like to meet the young man who has had the monumental audacity to attempt and succeed in jumping an electric wave across the Atlantic.'[19] To much laughter, the toastmaster went on to say that he himself had been chatting with Mr Edison a few days earlier: 'He said to me, "Martin, I'm glad he did it. That fellow's work puts him in my class. It's a good thing we caught him young."'

Marconi's modest response to this toast, and his generous acknowledgement of the work of his electrical predecessors, drew praise two days later from the *New York Times*, which also commented that for the proof of Marconi's achievement, 'nothing more was needed than Signor Marconi's unsupported and unverified statement. Immediately on receipt of telegraphic intelligence from Newfoundland that this feat had been accomplished and representative engineers of the world were interviewed, without exception their response was: "If Marconi says it is true, I believe it."'[20] The same report went on to say: 'He makes no boasts and indulges in no extravagant promises. He does not understand the art of promotion, perhaps, but he has established a character for truthfulness and conservatism, and

. . . we venture to say that he will have no need of the services of a promoter to capitalize on his invention.'[21]

Marconi was an astute businessman, and he certainly had no need of help to exploit his invention. He was determined to people the world with Marconi stations, all sending their messages over whatever looping paths it took to surmount Earth's curvature. He had confounded his mathematical critics. The only trouble was, he had no idea how.

Back in Britain, a certain mathematician heard the news and immediately guessed the truth. Oliver Heaviside was a striking man. Though not very tall,[22] he was arrestingly handsome, with a thick head of sandy-brown hair and sharp, disconcerting eyes. He was also very, very strange.

Tales of his oddities were legion. He furnished his rooms with blocks of granite; he dyed his hair black and then wore a tea cosy on his head until it was dry; he kept his nails exquisitely manicured and painted them cherry red. He lived most of his life as a recluse in a small Devonshire village, where the local adults regarded him with indulgence and the village louts catcalled after him and threw stones at his windows.

And yet, Heaviside's bizarre, sideways view of the world was also part of his genius. At school, in the same impoverished Camden slums that had produced Charles Dickens, Heaviside had railed against the conventional rote-learning that was all his teachers could imagine. To him, grammar was filled with 'unutterably dull and stupid and inefficient rules', and learning mathematics by mindless repetition had turned some of his fellows into 'conceited logic-choppers'.[23]

There was nothing of that in Heaviside's weird yet wonderful inner world. An editorial in *The Electrician* in 1903 wrote poetically of Heaviside's way of thinking: 'The ability to follow Mr Oliver Heaviside in his solitary voyages "on strange seas of thought" is given to few. Most of us do get but glimpses of him when he comes into some port of common understanding for such fresh practical provisions as are necessary for the prosecution of further theoretical investigation. These obtained, he steams fast to sea again. Some of us in our puny way paddle furiously after him for a little distance, but we are rapidly left astern, and, exhausted, laboriously find our way back to the land through the fog created of our own efforts.'[24]

Heaviside was impatient with people less bright than he – which included most of the population. He would sometimes try to make his work more intelligible, but he found it hard to believe that others couldn't grasp what seemed self-evident to him. 'In his most deliberate attempts at being elementary,' one scientist wrote, 'he jumps deep double fences and introduces short-cut expressions that are woeful stumbling blocks to the slow-paced mind of the average man.'[25] When a friend implored him to put a few extra words of explanation into a brilliant but fiendishly difficult piece of writing on electromagnetic theory, after his original 'it follows' he simply added '(by work)'.[26] Perhaps that was just an example of his impish humour. When another perplexed reader complained, 'You know, Mr Heaviside, your papers are very difficult indeed to read,' he replied, 'That may well be, but they were much more difficult to write.'[27]

Still, he could be charmingly self-deprecating. Even amid a ferocious 3,000-word diatribe against someone he believed had earned his censure, he wrote, 'I am full of nonsense, and, if it does not show, it is only because it is suppressed.' Heaviside was a prolific writer of letters, when indignant certainly, but also to amuse his friends. Perhaps surprisingly, his writing was always extremely neat, the words beautifully formed. Even his pages of mathematical equations were tidy, though they were occasionally enlivened with a quick sketch if he thought his formula suggested a funny face or figure. He also enjoyed playing with letters as well as numbers. He once signed himself: 'I am, Sir, Yours very truly and anagrammatically, O! He is a very Devil.'[28]

In spite of his reclusive tendencies, Heaviside craved visitors, but his strangeness and sarcasm frightened many of them away.

One scientist recalled his visit in 1914: 'I was very deeply impressed by Heaviside, even in my brief encounter. I never met anyone who, in spite of surface eccentricities, impressed me more deeply with the feeling that I was in contact with a really great mind. I have always been glad I made the visit, but I never had the courage to call again.'[29]

One of his best friends, electrical scientist George Searle, was unafraid of Heaviside's oddities and visited many times. The two of them talked of much more than science. Heaviside had become caught up in the recent craze for a new device called a bicycle. He and Searle would take their bicycles and go out 'scorching', a Victorian pastime that involved freewheeling perilously fast, to the imminent danger of any pedestrians. 'We used to put our feet on foot-rests on the front forks,' said Searle, 'and then let the cycle run down hill. Oliver put his feet up, folded his arms and let the thing rip down steep and quite rough lanes, leaving me far behind.'[30]

Once, realising that Heaviside needed spectacles, Searle insisted on finding a suitable pair. (Heaviside refused to go to the shop or even to listen to the possible options, but Searle found some anyway.) However, they had a difficult time of it. In the peculiar cod Latin that Heaviside often employed in his letters for humour, he later wrote to Searle and his wife: 'Georgio Searlio et Spousio. Salutem. Te igitur. Specs. Glass came out. Long hunt. Found accidentally in pocket.'[31]

Much of Heaviside's work involved complex theoretical aspects of the relationship between electricity and magnetism. He took the famous equations that had inspired Hertz's experimental work and reinvented them, changing the formulation so that they were infinitely simpler to manipulate. Even now, these same equations appear in textbooks in the form that Heaviside developed. He also proved that adding deliberate faults to telegraph wires could make them transmit messages more efficiently. This was a twisted side-effect of the way electricity and magnetism reinforce one another and was so counterintuitive that nobody would believe him. In the end, an American scientist used Heaviside's reasoning to file a patent. When American telegraph manufacturers began using this technique and showing how effective it was, the British eventually followed suit – but not before Heaviside had lost both the money and the credit for his invention.

When Heaviside heard about Marconi's achievement, he guessed immediately how the radio signals were travelling so far. He had already heard that short-distance wireless waves seemed to be bending beyond the horizon and had been fiddling a little with the problem. Some people thought the waves might actually be bending a little around the corners in a process called diffraction – the same way that a point of light seems to spread out if you half-close your eyes. Heaviside, though, was sure this wouldn't be enough.

But there was another option. Hertz had shown that something that conducts electricity, in his case a piece of metal, would reflect wireless waves in exactly the same way as a mirror reflects light. Thus, Heaviside said, there must be an electrical layer in the sky that acted as a sort of radio mirror, bouncing the signals back down to Earth so that they could defy the planet's curvature.

This isn't as strange as it might seem. All you need to conduct electricity is some electrically charged particles. The electricity that flows through wires into your house is made up of negatively charged electrons. But in principle, an electrical layer in the sky could be made up of either kind of charge, positive or negative. Or, more likely, both.

Although the air is extremely thin aloft, it still contains some floating atoms and molecules of gas. Every atom is made up of a small, very dense central nucleus, which is positively charged, and a floating cloud of orbiting particles called electrons, which are negatively charged.

Normally, these balance out exactly, and atoms and molecules are electrically neutral. But if something (for example, a cosmic ray slamming in from space) were to rip off a few electrons, it would leave behind a spray of positive and negative shards. In other words, the air would become electrical.

Though he never published the detailed maths, this reflecting mirror in the sky would come to be called the Heaviside layer.[32] (Since charged particles are called ions, we now call Heaviside's conducting layer of air the ionosphere.)

Heaviside's prediction of the ionosphere was one of the many important salvoes in a life's work filled with insights into electricity and telegraphy. But he always struggled for recognition and understanding, even when people managed to see through his difficult manner of expression to the genius that lay behind it.

While other people had made fortunes from patents based on his work, Heaviside was perpetually short of money, especially close to the end of his life. However, he couldn't bear anything that smacked of 'charity' and furiously refused a host of offers of help. When a friend brought him a loaf of bread, he was so enraged that he left the bread on display for a full year before another visitor insisted on throwing it away.

It didn't help his financial situation that Heaviside was thoroughly profligate with fuel. He had a horror of being cold. His room was usually 'hotter than hell', with both a blazing gas fire and an oil stove, and the windows kept tightly closed against any possible influx of refreshing air. This fear of cold extended to those around him. He had his housekeeper sign an agreement saying 'M W agrees to wear warm woollen underclothing and keep herself warm in winter.'[33]

Since he could rarely afford to pay for all this fuel, Heaviside had constant battles with the people he called the 'gas barbarians'. Toward the end of his life, unable to pay the bills, he was forced to go without gas for light or heat for nearly a year. A neighbour saw him sitting outside in his garden looking cold and ill. 'Go inside,' she said, 'and sit by your fire.' Heaviside smiled. 'Madam,' he replied, 'I have no fire – I have only my genius to keep me warm.'[34]

Apart from the gas, Heaviside didn't seem to care too much about material things. He also had very little patience with honours and awards. In recognition of his work on electromagnetism, he was shortlisted for the 1912 Nobel prize. He didn't win, but then again neither did the other illustrious people on the shortlist – including a certain Austrian physicist named Albert Einstein. All lost out to one Nils Gustaf Dalen, who had developed an automatic way of feeding fuel to lighthouses. Einstein, of course, went on to win the prize for physics a few years later, in 1921, but Heaviside didn't get another chance. Perhaps that's just as well, since it is hard to imagine him dressing himself up and going off to Sweden for the ceremony. On 4 June 1891, the Royal Society had tried to elect Heaviside as a Fellow. All he needed to do was present himself in London for the formal admission ceremony. Heaviside's response was a poem:

Yet one thing More
Before
Thou perfect be
Pay us three Poun'
Come up to Town
And then admitted Be
But if you *Won't*
Be Fellow, then *Don't.*

Of course Heaviside *didn't.* (But they made him a Fellow anyway.) Later he became even more eccentric and demanding about awards, turning them down or specifying strange conditions for accepting. Close to the end of his life, when the British Institute of Electrical Engineers wanted to award Heaviside their highest honour, the Faraday medal, they suggested sending a deputation to his house to present it in person. Heaviside was most upset. 'Who are they?' he wrote in great agitation. 'And I can't talk to more than one at a time, and that is not easy . . . and

I may not be able to get a room cleaned of the damcoal [*sic*] dust . . . Hadn't you better come one at a time on 4 successive days?'[35] When the news came that the institute had revised its plans and would send only one person with the medal, Heaviside was clearly relieved, and some of his famous impishness crept back into his reply: 'Very good . . . Alone, or with a lady to protect you against my notorious violence . . . I usually get on very well with ladies, with clear soprano voices that are so distinct and so unlike the throaty voices of gruff men. And they like me too, I think . . . though I don't flatter them . . . No Deputation. A lady for protection allowed.'[36]

Eventually, Heaviside couldn't continue alone in the house. (His housekeeper had moved out several years earlier, and nobody blamed her.) When he collapsed, Searle took him to a nursing home, where the nurses and other patients adored him. He died there on 3 February 1925.

Heaviside didn't know it, but his mirror in the sky would turn out to play a crucial role in protecting life on Earth. For now, though, even among physicists, his aerial mirror was simply friendly from the underside, bouncing Marconi's handy signals around the world and ending for ever the terrible isolation of ocean-going vessels.

**Sunday, 14 April 1912**

Harold Bride woke just before midnight. He lay in his bunk, listening to the crack-crack of the wireless operating key in the adjoining room. Instinctively, he translated the Morse code in his head. It was the usual passenger stuff, business arrangements, dinner party arrangements, see you soon, wish you were here. His friend and colleague Jack Phillips was obviously still working his way through the waiting mound of messages, now that the ship was within range of Newfoundland's wireless station at Cape Race.

More than a decade after Marconi's spectacular stunt at Signal Hill, every major passenger liner was equipped with one of his new wireless stations. They were staffed by boys from Marconi's own company, who could be distinguished from the regular crew by the Marconi emblem on their shiny jacket buttons, and on the fronts of their peaked caps.

Access to wireless was the *dernier cri* for luxury vessels, regarded by passengers as an engaging, expensive toy. Rich patrons used it to send their personal messages, or to keep abreast of the news during their long, luxurious passage across the ocean. Of course, wireless could be used to call for help, but few people took much comfort from this, or even regarded it particularly seriously.[37]

Still, wireless had made big, exciting news two years earlier when it enabled police to trap the notorious 'Dr' Crippen. Crippen's wife had been found murdered, bricked up in his house, her body partly decomposed with lime. A few days before the discovery, Crippen had absconded, taking with him his secretary, Ethel Le Neve. The case was a global sensation. Crippen's face stared out of newspapers the world over, sporting spectacles and a large drooping moustache.

A few weeks later, the captain of the *Montrose*, bound for Canada, had found himself growing curious about one of his passengers. This 'Mr Robinson' had shaved his moustache and was now growing a beard. He appeared to have the marks of glasses on the bridge of his nose, though he never wore any. He was travelling with his son, who seemed an unusually delicate youth, with trousers that were far too large for him and a hat stuffed with paper to make it fit. Though the youth was in his twenties, he still frequently held his father's hand.

Surreptitiously, the captain ordered the wireless operator to send a message to London. Inspector Dew, who was heading the Crippen investigation, immediately boarded the fast liner *Laurentic*, which would overtake the *Montrose* before it reached Canada. As 'Mr Robinson' remained sublimely unaware of the invisible messages crackling to and from the ship's aerial, newspapers printed daily reports and diagrams showing the positions of the two ships. The world watched the race unfold in front of them, and when Dew finally apprehended the two fugitives with the words 'Good morning, Dr Crippen, I am Inspector Dew of Scotland Yard. I have a warrant for your arrest', Marconi's wireless was the hero of the hour.[38]

The ship that Harold Bride was working, the mighty *Titanic*, had everything of the very best. Its wireless was the latest and greatest that money could buy. Pressing the communications key fired up the main condenser to a full ten thousand volts, and the leaping spark flung invisible waves hundreds, even thousands of miles, with a noise so deafening that the sending equipment had to be contained in a soundproofed room.

Bride's watch didn't begin officially for another two hours, but he knew that Phillips must be tired. Although wireless messages cost a princely twelve shillings and sixpence for the first ten words and nine pence per word thereafter,[39] the *Titanic* had plenty of passengers who weren't counting their pennies. It was to accommodate this preponderance of wealth that there were two operators on board instead of the usual one. Even so, they had lost seven hours of operation the day before to an annoying electrical malfunction, and since then both boys had been working overtime trying to clear the backlog of messages. Phillips had relieved the exhausted Bride half an hour early, and now Bride decided to repay the compliment. Still wearing his pyjamas, he pushed through the green curtain into the operations room.

Phillips was indeed weary. He wouldn't have needed much persuasion to yield his place. But before he could hand over to Bride, the captain put his head through the door. 'We've struck an iceberg,' he said calmly. 'And I'm having an inspection made to tell what it has done for us. You'd better get ready to send out a call for assistance. But don't send it until I tell you.'

The boys were mildly surprised; neither of them had felt a thing. They both waited at the set, and ten minutes later the captain was back. 'Send the call for assistance,' he said from outside the door. 'What call shall I send?' Phillips asked. 'The regulation international call for help. Just that,' was the response.

This was obviously more serious than it had seemed. Phillips immediately began to tap. 'CQD' he wrote, six times over, along with the call sign of the *Titanic* and its current position. CQD was the standard Marconi emergency signal, adopted in 1904. 'CQ', or 'seek you', meant 'attention all stations', and the added 'D' meant 'distress'. Two years later, the Berlin Radiotelegraphic Convention had recommended 'SOS' instead, which didn't mean anything but was a bit easier to recognise in Morse code.[40] Phillips had little truck with the new signal. He stuck to what he knew.

Inside the 'silent' room, giant sparks flashed and sent their mysterious invisible waves out into space bearing Phillips's cry for help. The time was 12:15 A.M.

Ten miles away, the *Californian* had hove to. Beset by ice, her captain had decided to wait until morning to continue. The lights of the *Titanic* were just visible in the distance, but nobody on board the *Californian* suspected any trouble. The ship's lone wireless operator had gone off duty at 11:30. He was already in bed.

Fifty-eight miles away, the *Carpathia*'s wireless operator, Harold Cottam, had also decided to go to bed. He was partly undressed when he remembered something that the boys on the *Titanic* might like to know. There was a tight network among the Marconi operators. Many of them knew each other personally, and they would often chat among themselves, ship to ship, when work was slow. For such conversations, you'd scarcely even need the ship's call sign. After a while, recognising someone's Morse touch was as easy as picking out a familiar voice in a crowd. You could tell from how quickly he pressed and released the key, from whether his touch was light or strong or hesitant, and sometimes just from little quirks that nobody else would spot.[41] The boys had unofficial shorthands amongst themselves. You could tell someone who was being annoying GTH, 'go to hell'. And to sign off, you'd say GNOM, for 'good night old man' (this in spite of the fact that they were all in their late teens or early twenties).

Cottam was a friend of both Phillips and Bride – in fact, he had recommended Bride for the job. Now he remembered that Cape Cod had some messages waiting for the *Titanic*. Perhaps he should let them know.

'I say, old man,' he tapped out, 'do you know there is a batch of messages waiting for you at Cape Cod?'

Cottam had been in his bunk-room for the *Titanic*'s first CQD, so he had no idea there was any problem. He was stunned by Phillips's immediate reply to his query:

'Come at once. We have struck a berg.'

'Shall I tell my captain?'

'It's a CQD old man. Position 41.46 N. 50.14 W. Come quick.'

On the *Californian*'s bridge, apprentice James Gibson idly studied the distant lights of the *Titanic* through his field glasses. At one point, he thought she was signalling with her Morse lamp. He tried to reply, but then decided the lamp was only flickering. At 12:45, second officer Herbert Stone of the *Californian* saw a warning rocket explode over the *Titanic* in a sudden flash of white light. How odd, he thought, that a ship should be firing rockets at night. Nobody on the *Californian* thought any more of the *Titanic*'s strange behaviour. The traditional methods of ship-to-ship communication had proved useless. Sight, after all, was blind.

But Heaviside's electrical mirror in the sky had already done its job. Even though the *Carpathia* was far over the horizon from the *Titanic*, the waves carrying Phillips's message had leapt over the intervening mountain of sea, before bouncing back down to where the *Carpathia*'s aerial crackled in response. Minutes after the *Carpathia*'s captain was wakened with the news, he ordered her to be turned and all power diverted to the engines. Cottam wired his friends on board the *Titanic* to say they were speeding to the rescue. They were four hours away, he wrote, and 'coming hard'.

Bride ran to tell the captain the news. When he returned, Phillips was sending more detailed directions to the *Carpathia.* 'Put your clothes on,' Phillips commanded. Until then Bride had forgotten he was still in his pyjamas. While Bride scrambled into his warmest clothes, an extra jacket and boots, Phillips never left the telegraph. He was firing out CQDs every few minutes, responding to any ship that replied, though most were hopelessly far away. He even tried a few SOSs, since, as Bride pointed out, it might be their last chance to use the new code. Meanwhile, Bride draped an overcoat over Phillips and strapped one of the *Titanic*'s distinctive white lifebelts to his back. They could both now feel the ship list forward. The water was up to the boat deck, and word came that the power would soon be gone.

At 1:45, the *Titanic* sent another message to the *Carpathia*: 'Come as quickly as you can old man; engine room filling up to the boilers.' That was the last message she received. A few minutes later, the captain appeared and formally released the two boys from their duties. From now on, he said, it's 'every man for himself'. The time was 2:00 A.M., and the lifeboats were all gone. Bride rushed to the bunk-room to get his and Phillips's money. As he returned, he saw a stoker who had sneaked into the wireless room and was silently slipping the lifebelt off Phillips's back. Bride was filled with rage. He recalled, 'I suddenly felt a passion not to let that man die a decent sailor's death. I wished he might have stretched rope or walked a plank. I did my duty. I hope I finished him. I don't know. We left him on the cabin floor of the wireless room, and he was not moving.'[42]

The band-members had given up hope of escape and stayed heroically at their posts, though they had switched from their insouciant ragtime music. As Bride ran to help some men struggling with a collapsible boat, which was lashed to the deck, he heard strains of the hymn 'Autumn', as if for a prayer: 'Hold me up in mighty waters, keep my eyes on things above.'[43] A wave took hold of the boat and washed it offshore. Bride found himself beneath it, shocked by the cold of the water, then somehow on top of it. The boat had overturned, and its occupants were now clinging to the waterlogged underside.

The night was eerily clear, brilliant with stars that reflected off the surrounding ice. There were no more sounds from the band. Also hanging from the upturned boat, seventeen-year-old passenger Jack Thayer was watching the ship with a horrible fascination:

> She was pivoting on a point just aft of amidships. Her stern was gradually rising into the air, seemingly in no hurry, just slowly and deliberately . . . Her deck was turned slightly towards us. We could see groups of almost fifteen hundred people still aboard, clinging in clusters or bunches, like swarming bees; only to fall in masses, pairs or singly, as the great after part of the ship, two hundred and fifty feet of it, rose into the sky, until it reached a sixty-five or seventy degree angle. Here it seemed to pause, and just hung, for what felt like minutes.

Then the lights went out. The ship's engineers had done their duty. They had stayed at their posts to feed electricity to the wireless set and power the waves that were spreading throughout the Atlantic, bearing Phillips's calls for help. Now every one of them was about to die.

The collapsible boat was so close to the ship that she was gradually being sucked back towards the great pivoting mass. Those who could crane their necks upward were aghast to see three huge propellers loom up over their heads. But then, the final intact bulkheads burst with a series of muffled thuds, and the *Titanic* slid gracefully and silently into the sea.

Jack Thayer heard nothing then but a single collective sigh. He recalled, 'Probably a minute passed with almost dead silence and quiet. Then an individual call for help, from here, from here;

gradually swelling into a composite volume of one long continuous wailing chant, from the fifteen hundred in the water all around us. It sounded like locusts on a mid-summer night, in the woods of Pennsylvania.'

For twenty, perhaps thirty minutes, these terrible cries continued, growing gradually fainter as, one after another, their owners succumbed to the cold. All the while, half-empty lifeboats stood off silently, only a few hundred yards away, their huddled occupants afraid to take anyone on board lest their boats be swamped.

The upturned collapsible boat was now little more than a raft, and yet those already clinging to its keel heaved up whoever else they could until the boat was so low in the water that there was no room for more. Now there were twenty-eight people on board, standing, sitting, lying, kneeling, crammed together any which way with no hope of shifting. Someone was kneeling on Thayer, holding his shoulders, and somebody else was on top of them both. Bride was lying full length, his feet crushed against the cork fender, which was under two feet of frigid water; someone else was sitting on his legs. For the next two hours they clung thus, the only cheer Bride's recurring reassurance: 'The *Carpathia* is coming up as fast as she can,' he told them again and again. 'I gave her our position. There is no mistake. We should see her lights at about four or a little after.' Though Bride couldn't see him in the darkness, Phillips, too, was crammed on to the boat. But he stayed strangely silent.

The *Carpathia* did come; her lights appeared as Bride had predicted, a little after four. All aboard the collapsible boat were saved, apart from one stiff figure who, it turned out, had already succumbed to hypothermia. In spite of Bride's best efforts to drape him with clothes, Phillips hadn't taken the time away from his messages to dress himself warmly enough for the Atlantic's frigid waters. He never had a chance.

Bride had to be carried aboard the *Carpathia*, his feet too crushed and frostbitten to bear him. Still, after a few hours in the ship's hospital, he joined Cottam in the wireless room, where he remained, sending passengers' messages of grief, until the ship reached New York.

He was still at his station, clicking away, even after the ship had docked, when the door opened and a voice said, 'Hardly worth sending now, boy.' 'But these poor people, they expect their messages to go,' Bride replied. Then he turned and realised who was in front of him. A lowly ship's operator would never have met Mr Marconi personally, but every wireless room bore his portrait. Bride looked up at the picture on the wall, and then across again to where Marconi was standing. Marconi extended his hand, and Bride shook it without a word. Then he tried to smile and failed. 'You know, Mr Marconi, Phillips is dead,' he said.[44]

In all, more than 1,500 souls had perished. Marconi, who had arrived in New York just a few days earlier, was terribly shaken by the disaster. He had originally planned to travel on the *Titanic*, but he had been so far behind on his paperwork that he had changed his passage to the *Lusitania*, which had an extremely competent stenographer on board. Still, his wife Bea and two young children should have been on board the *Titanic*, to meet him in New York for a holiday. Instead, his son Giulio had fallen ill and his wife had cabled that she was postponing the trip.[45]

Despite the enormous losses, 712 people had been saved, all through the combined power of wireless waves and their horizon-beating mirror in the sky. There were criticisms of Marconi, of course. Had he ordered his operators on board the *Carpathia* to withhold their news until he could sell their stories more effectively? Bride and Cottam did both sell their stories to the *New York Times* for what must have been colossal sums to them, amounting to three or four times their annual salaries.[46] A later US Senate investigation reported: 'Some things are dearer than life itself. The refusal of Phillips and Bride, the wireless operators, to desert their posts of duty even after water had mounted to the upper deck, is an example of faithfulness worthy of the highest praise.' But Bride couldn't shake off the memory of that stoker he had killed, and he kept changing his story. He and Phillips had struggled with the man together. Then it was Phillips alone who had done the killing. Bride returned to England a hero, but shortly before the tenth anniversary of the disaster he disappeared to Scotland, changed his name and became a travelling salesman. He had his own radio set still, and chatted occasionally over the airwaves to people who had no idea of his story.

But the fact remained that if Marconi's waves had not bounced their way over the curving ocean, nobody would ever have known the fate of the *Titanic*, and everyone on board would have been lost.

Now that the power of wireless had been amply demonstrated, everybody wanted one. Wireless stations sprang up around the world. Marconi made vast amounts of money and acquired all the fame he could ever have wanted. He even won the Nobel prize for his invention. However, in spite of Heaviside's prediction nobody, least of all Marconi, knew what had created that mirror in the sky.

The wireless world needed another physicist, someone who, like Oliver Heaviside, could understand the mysterious workings of Heinrich Hertz's rays. The person it got was measured, cool, precise, diligent and utterly conventional, in fact everything that poor Heaviside was not.

Edward Victor Appleton ('Vic' to his family) was born in Bradford, in the north of England, in 1892. His family was working-class. They lived in a typical grim neighbourhood of back-to-back houses, dominated by the industrial outpourings of the local mill. But though Appleton's neighbourhood was poor, it was also respectable. The curtains that shielded the windows were spotless. Even the stone steps that fronted these small houses were always meticulously scrubbed. Appleton's father, a warehouseman, wore a bowler hat every day instead of the plebeian flat cap of the mill-workers. And his neighbours were policemen, railwaymen and postmen, who wore their uniforms as badges of their reputability.

Appleton was a golden boy. At the age of eleven he won a scholarship to a first-rate high school, where he excelled at everything he touched. He had a fine singing voice; he was captain of the football and cricket teams; he was handsome and popular, with grave grey eyes and wavy brown hair that the girls loved; he ranked top in just about every academic subject from literature to science; and he was the only pupil ever allowed a key to the physics laboratory, so he could continue his work there in the evenings. Aged eighteen, he won another scholarship, this time to Cambridge University. To help set him up there, his parents cashed in an endowment policy and his uncle gave him a present of five gold guineas.

In some ways, Cambridge suited Appleton perfectly. When he arrived as an undergraduate at St John's College in 1911, his conservatism was already engrained. He wore stiff collars made by a Bradford tailor; he would continue to buy the same collars, from the same shop, for the rest of his life. He was also dazzled by the splendours of St John's, one of the oldest and richest colleges in Cambridge. Almost immediately, he dispatched a postcard to a Bradford friend, saying, 'I have fine rooms of my own, and can feel my importance, I can assure you.' The picture on the front of the card showed a dining hall from one of the other Cambridge colleges, and Appleton added as a PS: 'We at St John's have a dining Hall even bigger and finer than this.'[47] More impressed postcards followed. 'I went to have breakfast with a man I know – a tutor of Clare College,' said another. 'He had some of the College silver plate on his breakfast table. Being a tutor, the College lends it to him now and then. Jove it was a gorgeous sight.'

Appleton excelled at Cambridge both athletically and academically, and he graduated in June 1914 with a dazzling double first in physics. Two months later Britain declared war on Germany, and Appleton immediately joined up. But when the war was over he returned to St John's as a fellow. He was still entranced by the Cambridge traditions: the gowns and formal portraits, the silver and gold plate,[48] the flickering candles and Latin graces in the dining hall. But while part of him was glorying in the unexpected free pass he had received into this august world, Appleton remained enough of a Bradford outsider to begin to notice some of its downsides.

For instance, when Appleton asked that the cockroaches in the college kitchen be disposed of, he was astonished by both the steward's refusal and the reason he gave. The St John's cockroaches, it turned out, had been brought over from the Continent sometime during the reign of Elizabeth I, and were not to be disturbed. And although Appleton relished the approval of his clever, confident peers, he didn't much like their snooty attitude towards the world outside the university walls. Later, he often mentioned a Cambridge don who boasted that he had never been inside a cinema: 'He made the remark in a way which showed he expected admiration all round, and the sad thing is that he got it – nearly all round.'[49]

In that way, at least, Appleton would never fit in at the Cambridge of the time. For, formal as he was, and as carefully as he had shaken off the appearances and accent of his working-class youth, he liked putting people at their ease. Years later, when he was chancellor of Edinburgh University, he would reach up and dust the parts of his filing cabinet that his diminutive cleaning lady couldn't reach. He would chat to servants about soccer, or anything else he could think of that they might like. And when he was meeting with his similarly illustrious colleagues and timid secretaries entered the room with tea, he would tease them with the declaration, 'Here comes Hebe – cupbearer to the gods!'[50]

He also talked about his scientific work to anybody who was curious enough to listen. That meant not just academics but ordinary people, the sort that many other dons disdained. He gave public lectures that were clear and entertaining, carefully practised beforehand and then modulated by his fine tenor voice. His main motivation sprang from the exuberance he felt in finding out something unexpected and magical about the world. For Appleton, science was all about imagination.[51] Years later, in a presidential address to the British Association, he would say:

> Perhaps the most striking fact about modern science is that, like poetry, like philosophy, it reveals depths and mysteries beyond – and, this is important, quite different from – the ordinary matter-of-fact world we are used to. Science has given back to the universe . . . that quality of inexhaustible richness and unexpectedness and wonder which at one time it seemed to have taken away.[52]

And as Appleton performed his research in Cambridge in the early 1920s, he realised he was on to something that was very far from the ordinary, matter-of-fact world. Working in the Signal Service during the war, he had become intrigued by the new radio technology, especially devices called thermionic valves. These were so crucial for signalling that they were classified as a military secret. However, using them was an inefficient process of trial and error, since nobody seemed to understand how they worked. When Appleton returned to Cambridge after the war, he had with him several of these mysterious valves ('not I may add lifted from the British Army. Some were gifts from the lamp firms who had been making valves, and one German type was picked up by me from a captured pill box').[53] Using them, he began to pick apart how exactly this wireless technology of sending and receiving radio waves really worked.

And the more he studied Marconi's radio waves, the more Appleton became intrigued by how they managed to curve their way around the world. He had met Marconi and reported being most impressed by the way he never let a discouraging theory stand in the way of an experiment. Even now, more than twenty years after Marconi had bent his waves over the Atlantic, there was still much confusion about exactly what was causing them to bounce.

Appleton thought the likeliest explanation was Oliver Heaviside's idea: somewhere high overhead, the air crackles with electrical energy, which acts like a mirror bouncing the radio waves back down to Earth. But he wanted to go further. What exactly was this mirror like? What formed it? And how did it work?

He believed that a clue lay in something the Marconi wireless boys had known for years: certain times of day were better than others for sending wireless signals. As one commentator put it: 'Every operator has experiences of his own to tell when all the elements seemed to unite in his favour and the mysterious spark has travelled for almost inconceivable distances.'[54] And the best time of all seemed to be at night. Nobody knew exactly why this should be, though some speculated that perhaps there was less interference between messages at night, when fewer stations were active. Appleton, though, had a better explanation. He thought the night/day difference meant that the sun had something to do with making Heaviside's electrical layer.

Perhaps something that arrived with the sun's rays somehow split the constituents of the upper atmosphere into their electrical pieces, ripping electrons off the floating atoms and molecules, and splattering positively and negatively charged shrapnel around the sky. That would certainly fill the sky with electricity. It would also happen in the outermost atmosphere, our first aerial rampart against invaders from space.

But if this was right, why should the layer reflect radio more efficiently at night, when the sun had vanished? Appleton thought he understood. He reasoned that at least remnants of the electrical layer would be there all the time, even when the sun had vanished. Because positive and negative charges attract each other, the electrical shrapnel will gradually recombine. But the uppermost air is thin, pieces don't encounter each other very often, and one night would not be long enough to lose everything.

However, the night-time layer would be different from the daytime one in this crucial respect: it would be higher in the sky. During the day, the sun's rays, or particles, or whatever was responsible, would be able to penetrate deep into the atmosphere. Heaviside's reflecting layer would reach down into the part of the sky where the air is relatively dense. Any radio waves arriving at this low, dense electrical mirror would not only bounce off it; they would also collide and be partially absorbed, and would lose some of their energy in the process. At night, this low-lying layer would thin out and rise as the electrical shrapnel recombined. All that would be left were the remnants of electricity in the high, thin air where collisions were rare, and recombination slow. There, radio waves would be able to bounce off more efficiently, without losing so much energy. Hence radio would travel farther at night.

In April 1924, Appleton took on an assistant, Miles Barnett, from New Zealand, to test his ideas. Barnett immediately set to work measuring the radio signal arriving in Cambridge from London. For wireless had come a long way from Marconi's initial dot-dot-dot across the ocean. Now its waves floated out accompanied by voices, and even music. Two years earlier, the newly formed British Broadcasting Company had started a station called 2LO in London, and their regular broadcasts could be picked up in Cambridge. Appleton asked Barnett to measure how strong this signal was at different times of day and night.

Barnett had no trouble picking up the 2LO signal, and he quickly confirmed that it was stronger at night than during the day. But he also spotted something curious. Every day, around dusk, the 2LO signal definitely wobbled. It faded into and out of existence, as if some cosmic sprite were fiddling with the volume. Both Appleton and Barnett realised immediately what this meant. The signal had to be tracking Heaviside's electrical layer as it made its nightly move upwards.

The signal Barnett was measuring had two components: a wave that came to him directly, and another that bounced off the mirror in the sky. When these two waves arrived at his Cambridge transmitter, they had the potential to interfere with one another. If the difference between the two paths amounted to a whole number of wavelengths, the two waves would combine to make a superwave and the 'volume' would lurch upward. If, on the other hand, one wave had travelled exactly half a wavelength farther than the other, it would hit its peak while the other still languished in its trough. Then the two waves would cancel each other out, and the signal would flatline.

It would be a tremendous coincidence if either of these things happened normally. The wavelength of the wireless waves was set by the BBC, and the distance they travelled by a combination of the location of Appleton and Barnett's lab and the height of Heaviside's layer. There was no earthly reason why these different random figures should conspire to make the waves neatly add up or subtract from one another. And normally, in full day or full night, they didn't.

However, at dusk something changed. The Heaviside layer began its nightly foray upwards into the sky. As it did so, it passed through the exact heights that would make the reflected radio wave interfere with the one on the ground. Imagine the layer as it gradually rises. First, it reaches a point where the reflected wave will exactly match the ground wave, peak for peak, and up goes the signal volume. The layer keeps on rising. Now it passes through the perfect height for the reflected radio wave to cancel out its ground-based twin. Suddenly, the signal drops to zero. The layer rises farther and hits another mutual peak, followed by another cancel point. Loud, soft, loud, soft – as the Heaviside layer rises, the signal in Barnett's measuring apparatus wobbles in volumetric sympathy. It was the first direct evidence that Heaviside had been right.

This gave Appleton an idea. Obviously he couldn't change the height of the Heaviside layer for his experiment, but he could, if he asked nicely, get the BBC to change their wavelength. If they could be prevailed upon to broadcast a gradually changing signal, that should have the same effect as the rising electrical layer – the waves would gradually move through points where they added together and subtracted from another, and the ultimate signal should have the same wobbles of interference. That would confirm the Heaviside layer was really there. But more than that, it would reveal exactly how high this mysterious mirror lay.

That's because Appleton could count the highs as he moved through them. Each one would show him the point where the new wavelengths added up. Knowing this, the broadcast wavelength, and the distance from the broadcasting station to his lab, he could work out how high the reflected wave had to go before it bounced back and arrived again at his apparatus.

It was a brilliant idea, and the BBC quickly agreed. Station 2LO couldn't easily broadcast a gradually changing wavelength, but they could do it instead from Bournemouth, on the south coast. That meant recalculating the appropriate distance from station to lab, and to Appleton's chagrin he realised he would have to move his experiment from his beloved Cambridge to a borrowed lab in – of all places – its sworn rival, Oxford University.

On 11 December 1924, Appleton and Barnett set up their equipment.[55] They waited impatiently for Bournemouth to finish the regular broadcast. To Barnett, the last number, by the Savoy Orpheans, seemed to go on for ever. 'And I always thought I liked dance music,' he groaned. Finally, just before midnight, the programme ended. On the telephone Captain West, head of the Bournemouth station, told the two researchers to be ready. And then, a few minutes after midnight, the changing signal began. A few minutes later came the wobbles that Appleton had been looking for. Heaviside's crackling layer of earthly electricity was floating some 100 kilometres above his head.

Appleton had discovered what Heaviside had only imagined. Now the real work was to begin. From his new position as head of the physics department at the University of London, Appleton set up a network of researchers to study the new layer. The task of broadcasting ever-changing signals was switched to the National Physical Laboratory at Teddington, and Appleton set up various new receiving stations including two wooden huts, which were erected just outside Peterborough.

To run the Peterborough site, Appleton hired a new assistant, one Mr W. C. Brown, who had been a shipboard radio operator in the war and had travelled very widely. Among other things, this had made him highly resourceful. 'He could produce a cup of hot tea, in the middle of the night, when there was neither tea, nor milk, nor cups,' Appleton said. 'And when Mrs Brown joined him, as she did from time to time, the most delicious sausage rolls also used to appear, again from nowhere. All the early work on the ionosphere was done on tea and sausage rolls.'[56]

The first use of the Peterborough side was to test what happened at dawn. Using the National Physical Laboratory, Appleton had much more flexibility about when his test signal could be broadcast. Since he already knew that the Heaviside layer rose with the dusk, he wanted to check that it fell again with the dawn. As he expected, as the sun's rays returned to electrify the atmosphere, the reflection of the radio waves grew steadily fainter and the layer dropped – in some cases to just fifty kilometres or so.

But this still left the intriguing question: how exactly was the sun achieving this? Appleton wanted to know what made the Heaviside layer.

On 29 June 1927, he got his chance to find out. For on that day a solar eclipse was to take place, with the moon passing in front of the sun and blocking its rays from Earth's view. The moment the sun's rays were stopped, would it be like dusk? And when they returned, would the Heaviside layer deepen as if at dawn?

The eclipse would take place early, about five in the morning. Appleton persuaded the ever-obliging BBC to send out special transmissions from Birmingham for him to pick up at Peterborough. He also talked ships' captains into limiting their transmissions to emergencies for the duration of his experiment, to keep the airwaves free. Sunrise came on the 29th, and with it the familiar lowering of the Heaviside layer as the sun's rays slowly appeared. Then the time for the eclipse approached. The moon's shadow began to shroud the atmosphere above Peterborough, and immediately the reflected radio wave grew stronger and the Heaviside layer lurched upwards.

The effects at dawn and dusk had always been gradual, reflecting almost imperceptible changes as the sun's rays gradually spilled over the horizon. But now, with this solar eclipse, there was no delay. The effect was instantaneous.

That, to Appleton, could mean only one thing. Whatever was doing the electrification had to travel earthwards at the speed of light. No particles can travel that fast – it had to be some kind of light rays. And Appleton correctly surmised that the culprit must be among the most energetic such rays in the universe: x-rays.

This was the answer he had been looking for, the explanation for why Heaviside's mirror was present in the sky. But more than that, it revealed for the first time the powerful role that this mirror plays in all our lives. Because the same process that electrifies the air also protects us from a horrifying menace.

X-rays from space are deadly, because they can do to living cells precisely what they do to the atoms high in the ionosphere. Incoming x-rays arrive with such extraordinary energy that they can smash DNA molecules into their electrical fragments, triggering the onset of cancer. That's why we have to be so careful about medical x-ray doses.

X-rays are so energetic that it takes only a smidgeon to allow us to see through biological tissue and pick out the body's internal organs and bones. So there's not much to worry about in a hospital – it would take 45,000 simultaneous chest x-rays to kill you. But outside the ionosphere's protection, that kind of blast can happen in an instant. The sun throws them out all the time. Just one solar outburst of x-rays, and any creature that is not sheltering beneath the ionosphere would be fried. The

International Space Station contains a specially reinforced module to protect the inhabitants from this danger. If the sun flares up, astronauts must immediately rush to the module and take cover.

With his eclipse experiment, Appleton had discovered that the ionosphere was much more important than a mere conduit for communications. It forms yet another sacrificial layer of air. By allowing its atoms to be shattered, it protects us from the x-rays that continually bombard our fragile planet.[57]

Appleton's fame was now assured, and the conventional high honours came his way. He was knighted by the king. He received the US Medal of Merit, the Officership of the French Legion of Honour, and was even appointed by the pope to the Pontifical Academy of Science. Like Marconi, and unlike Oliver Heaviside, Appleton also received the Nobel prize for physics.[58] In his later years, Appleton became an illustrious university administrator, spending much more time in committees than he managed in the lab. He became more Establishment than ever, but his conservatism still came with a twist. His younger daughter, Rosalind, exhibited a 'slight naughtiness' that made her a favourite with her father. He was delighted when, dissatisfied with a drink served her at a hotel, she calmly topped it up from a bottle of gin that she had withdrawn from her straw shopping basket.

And whenever he could, Appleton would sneak away to collect data or analyse results. He talked about his research as 'escape into the upper atmosphere'.[59] He spent the rest of his life trying to understand how the ionosphere worked.

One of his most intriguing, and baffling, findings had come back in the 1930s during an expedition to the Arctic. Appleton had wanted to investigate how the ionosphere was affected by magnetic disturbances, which were known to be strongest near the poles. So he arranged to take measurements from Tromsø, in the far north of Norway. While he was there, Appleton discovered that there seemed to be a connection between the ionosphere and magnetic storms. When a storm arrived to set compass needles swinging, the ionosphere blacked out. Electricity and magnetism were obviously working in some kind of tandem.

Though Appleton didn't figure out exactly how these two forces co-operate in the air above our heads, he was certainly on the right track. For the final protective layer of the atmosphere is indeed driven by a co operative effort between the electricity of the ionosphere and the magnetism above it. Thousands of kilometres above Earth's surface, where the air is so thin it is scarcely there at all, sweeping lines of force from the planet's own magnetic field act as sentinels for one final threat from space, while below the ionosphere is waiting to catch this threat and disarm it.

This threat is the most dangerous of all, and yet we were unaware of it until the 1950s, at the very dawn of the space age.

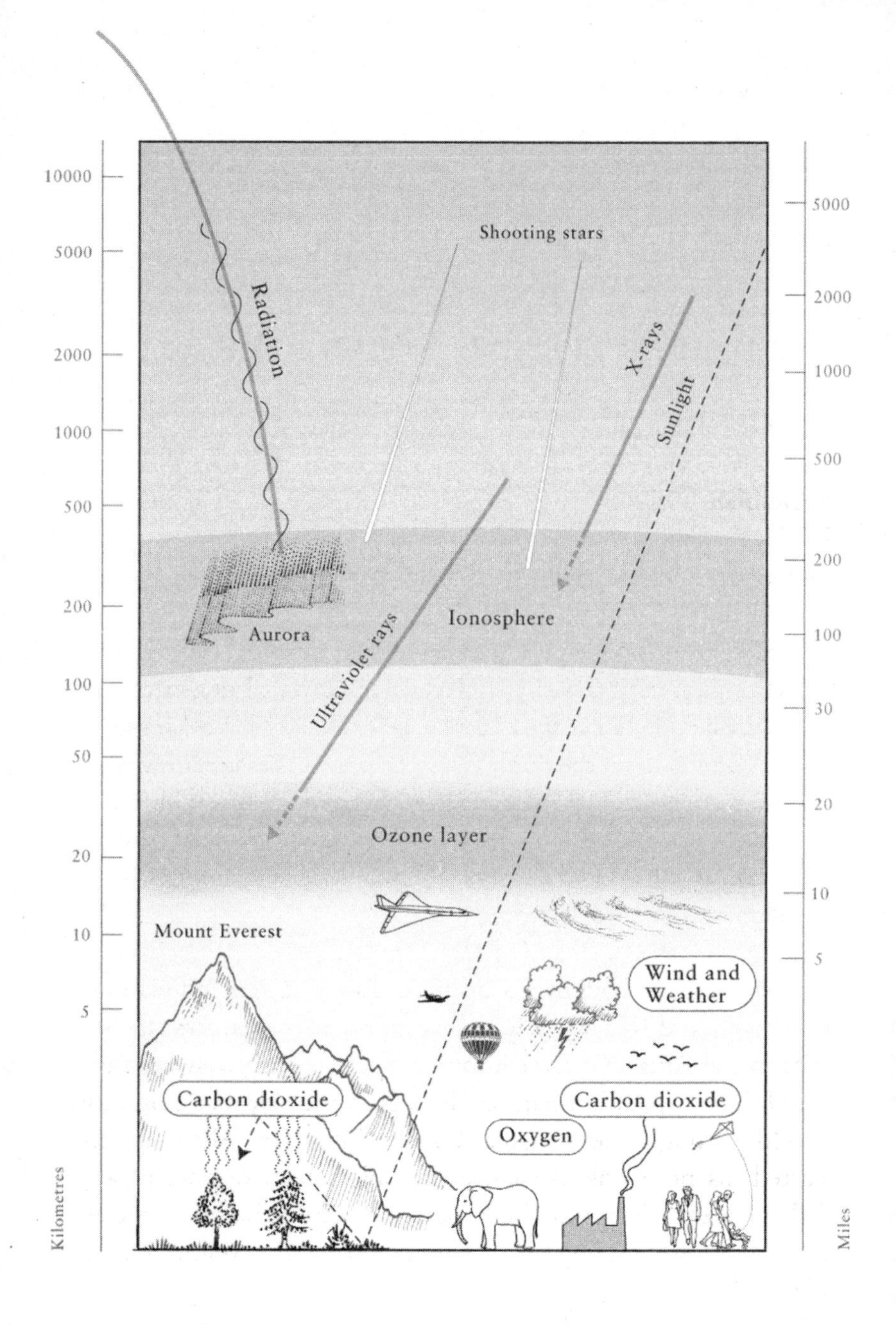
10000
5000
2000
1000
500
200
100
50
20
10
5
Kilometres
5000
2000
1000
500
200
100
30
20
10
5
Miles
Shooting stars
Radiation
X-rays
Sunlight
Aurora
Ionosphere
Ultraviolet rays
Ozone layer
Mount Everest
Wind and Weather
Carbon dioxide
Carbon dioxide
Oxygen

## *Chapter 7*

# THE FINAL FRONTIER

**4 October 1957**
**USS *Glacier***
**Somewhere in the vicinity of the Galápagos Islands**

> Yesterday night the 4th and early this morning were very exciting for me (as well as for the civilised world in general).
>
> Just before dinner time Larry Cahill told me that news was just coming in on the ship's news circuit that the *Soviet Union had successfully launched a satellite.* Factual details as follows:
>
> Inclination of orbit 65 degrees to earth's equator. Diameter 58 cm, Weight 83.6 kilograms (Wow!), Estimated height 900 kilometres. Period 1h 35m

James van Allen had always been assiduous at keeping his field notebook, and this one was no exception. It was neatly labelled 'Equatorial-Antarctic Expedition', and every entry was carefully dated. The expedition had barely begun; the ship had only just passed through the Panama Canal. Still, Van Allen usually started his notebook as soon as he sailed, to take note of any little incidents that might be important later. He hadn't expected to be writing anything as exciting as this, and certainly not so soon.

Van Allen ate his dinner. He watched a second-rate movie. But he couldn't contain himself; even out here in the middle of the ocean, he needed to know more. Off he headed to the communications shack, where a young radioman was already seated, wearing a set of headphones and hovering over a receiver. 'I think I have it,' he said. Van Allen took the headphones and listened for himself. There it was, loud and clear: 'Beep-Beep-Beep.' It was incredible to believe that this was an artificial satellite, launched by human hand and ingenuity, announcing its presence as it passed sporadically above the ship with this regular, disciplined cheeping. It was so unlike the natural – erratic – sounds of the atmosphere, and so exactly what Van Allen had been wanting to hear for years.

He had been saying since 1948 that humankind could put a satellite into orbit. The *New York Times* had mocked him; *The New Yorker*, even, had poked its own brand of gentle fun. He had been forced to remove that part of his speech at a major conference as being 'too speculative'. And now there it was, Beep-Beep-Beep, right above their heads.

Immediately, Van Allen wanted a recording. But the tape recorder down in his lab was way too bulky, and besides, it was built so completely into the other apparatus that it would take too long to free. By now another passenger was in the communications room – John Gniewek, from the US Coast and Geodetic Survey, who would be operating a magnetometer station in Antarctica for the next year. Gniewek had a small magnetic tape recorder in his room – he could easily fetch it, he said. While Gniewek did this, Van Allen tumbled downstairs to fetch his oscilloscope. What would it look like, this first human signal coming down to us from space? Van Allen noted the shape down in his notebook: a flat line punctuated by a periodic thicket of scribbles, as if an unruly child were grabbing the pencil every 0.2 seconds, for precisely 0.3 seconds at a time.

The communications shack was getting crowded now as more members of the expedition arrived. They took turns listening to the next pass, and the one after, and the one after that. Finally, at two in the morning, Van Allen headed off to bed. He had rarely written so many exclamation points in his field notebook. And the last one for the day was perhaps the most significant: 'Very great thrill!'

'The dawn of the space age!' screamed newspaper headlines around the world. London's *Daily Mirror* changed its header. No longer the 'biggest daily sale in the world', it had become the biggest 'in the UNIVERSE'.

News had come quickly to Washington. As luck, or, more probably, design, would have it, scientists from the Soviet Union, the United States and five other nations had gathered at the National Academy of Sciences to discuss rocket and satellite activities for the International Geophysical Year, which was currently under way. A member of the Soviet delegation, Sergei M. Poloskov, had already caused jitters by suggesting that the world was on the eve of the first artificial earth satellite. And now, there it was. Walter Sullivan from the *New York Times* had received a call from his desk editor. Immediately he had rushed over to one of the US attendees. 'It's up!' he had whispered. The attendee threaded through the crowd to relay the news to the official US delegate for the meeting, one Lloyd Berkner. Berkner called for silence. 'I wish to make an announcement,' he said. 'I've just been informed by the *New York Times* that a Russian satellite is in orbit at an elevation of 900 kilometres. I wish to congratulate our Soviet colleagues on their achievement.'

America, of course, was thunderstruck. At first there was silence, and then the jokes came, followed quickly by recriminations. Bars around the country began selling 'Sputnik cocktails' – one-third vodka and two-thirds sour grapes. And everybody wanted to know how in the hell the Russians had got there first. The United States was the country of technological innovation. It was the country that had pioneered flight and had been leading the world for decades. How had the American satellite programme been caught, as one sour punter remarked, 'with its antennas down'?

There was no shortage of theories. Coming as it did on the coat-tails of the McCarthy era, some said the witch-hunting of scientists was the problem. Others laid the blame higher. Hadn't the president himself talked repeatedly of scientists as 'just another pressure group'? Hadn't presidential aide Sherman Adams talked disparagingly of 'an outer space basketball game'? There was only one thing that everyone was clear about: Americans needed a riposte, and they needed it fast.

Vanguard was the name of the official US satellite programme. After a dreadful summer of technical hitches, the project team was about to get its first full rocket off the ground. But the upper two stages of the rocket were to be dummies. And nobody was interested any more in tests. What they needed was a *satellite*.

The programme's director, John Hagan, did his best to explain to the president the current state of play. They had another launch scheduled for December that year, and yes, this time it would be a complete rocket, no dummy stages. It would also carry a minimal payload – a satellite, if you like, weighing two kilos. However, this was not, repeat *not*, a mission flight. It was designed merely to test the launch vehicle. Putting the satellite into space from this flight would, said Hagan, be 'a bonus'.

On 9 October, the presidential press office informed reporters that, in two months' time, Project Vanguard would launch a 'satellite-bearing vehicle'.

Now the pressure really began to rise. Early in November, the Vanguard test rocket, known as TV-3, made its way to launch complex 18A at Cape Canaveral in Florida. For the next four weeks, all tests went smoothly. The engineers were cautiously optimistic, even though the crowds who had started to arrive at the Cape were making them uneasy. This was supposed to be a *test*, done under carefully controlled conditions, with a bit of peace and quiet. But the president's announcement had put paid to that. Word had got around that this was America's attempt to put a satellite in space. And everybody wanted to be a witness.

Or almost everybody. Hagan had decided to stay in

Washington to keep an eye on events there. His deputy, Paul Walsh, would let him know exactly what was happening on the ground.

The *New York Times* was at the scene, of course. 'Last night,' its reporter wrote on Sunday, 1 December, 'from one of the coarse sandy beaches where the "bird watchers" of the missile age watch the Cape Canaveral spectacles, the Vanguard Tower was clear against a starry sky, two bright white lights glaring at its base and a red beacon shining at its top.' It was certainly an impressive sight: the white rocket, nudged up against its giant gantry crane, pointing to the sky. People had arrived from all over America, and even from Europe. As the days advanced, the excitement of the crowds only mounted as one delay after another halted the countdown clock. Launch had been scheduled for Wednesday, then Thursday. But finally, by the morning of Friday, 6 December at 10:30, there were only sixty minutes to go.

*T-45.* The radio tracking network began sending in its 'all clear' signals. *T-30,* and a klaxon sounded warning all non-essential personnel to leave the area. *T-25*, and the heavy doors of the blockhouse swung shut. *T-19,* the blockhouse lights were now out. *T-5*, and the voice reading the countdown had the faintest of tremors. *T-1*, and the count had switched to seconds. The rocket fired, the engines lit with an unspeakable roar. And . . .

'Look out! Oh god no!' 'Duck!' Most people in the control room did indeed duck. The rocket had collapsed into a spectacular ball of flame. (Unremarked for the moment, the satellite itself had tumbled from the nose-cone and lay beeping on the ground, alive, but hopelessly dented.) From his vantage point northwest of the control room, Paul Walsh had been counting down on the phone to Hagan back in DC. 'Zero, fire, first ignition,' he had said. Then: 'Explosion!' 'Nuts,' was Hagan's reply.

It was clearly time for Plan B. The army, who had been privately working on their own rocket-delivery system for years now, stepped into the open. Back in October when Sputnik went up, Neil H. McElroy, the new secretary of defence, had been making a tour of military installations around the country prior to taking up office. On 4 October he was at Redstone Arsenal when the news broke. Army rocket scientist Wernher von Braun had always wanted his project to be picked, instead of the Navy-inspired one that became Vanguard. Now he was almost tearful in his pleas: 'We knew they were going to do it. Vanguard will never make it. We have the hardware on the shelf. For God's sake turn us loose and let us do something. We can put up a satellite in sixty days, Mr McElroy! Just give us the green light and sixty days.'[1]

A discreet green light had – eventually – winked at Von Braun from the President's office, and he was now more than ready to step in. He didn't only have the rocket hardware good to go; he also had a satellite. Because one of the eager scientists who had been preparing payloads for possible launch in the International Geophysical Year also had the farsightedness to design his machine to fit both Vanguard *and* the army's rocket contender, the Jupiter C. That scientist was currently on a ship out in the Pacific Ocean, and his name was James A. van Allen.

Van Allen received the first Marconi radiogram while he was still on board the *Glacier*. It arrived on 30 October and read: 'To Dr Van Allen, Would you approve transfer of your experiment to us with two copies in spring. Please advise immediately.'

He wasn't surprised. The day after Sputnik was launched, he had written a series of comments in his field notebook, under the previous night's jottings. 'Brilliant achievement!' was the first, followed swiftly by '?? Where do we stand now on Vanguard?'

Hastily he wired back his assent. Yes, the instruments could be transferred to the Jet Propulsion Laboratory with his very good will, to be readied for a launch in the spring. The *Glacier* had almost completed its mission now, and by early November was steaming into Lyttleton Harbour, New Zealand. Van Allen gathered his bags and hurried home to Iowa.

Van Allen had been delighted when the job came up at the University of Iowa. He had been born and raised in the state and was happy to go back after his stint in the east, at Johns Hopkins. His wife Abigail hadn't been so sure. She was an easterner; he'd met her in Baltimore – bumped into her quite literally while he was driving one day to his lab. (It was at a stop sign. The damage to the cars was minor, but the effect on the drivers was rather longer-lasting, since they got married six months later.)

Abigail had been west of the Mississippi only once, and was convinced that she might as well go to the moon as far as culture was concerned. The family had arrived on a frigid New Year's Day seven years earlier in an old station wagon, pulling an even older trailer. Van Allen himself admitted that their first tiny apartment there 'presented a challenging problem in heat transfer'. But things were better now. The family was settled, Abigail was happy, and the work had been going well even before the satellite world had heated up so dramatically.

Trained as a physicist, Van Allen had been a naval officer during the war, but it was afterwards, with the chance to experiment with captured German V-2 rockets, that his scientific sights soared. From then on, he could only look up, to the outermost reaches of Earth's air. He wanted to know what filled these tenuous threads. Perhaps there were particles from outer space that pinged periodically on to the planet's outer skin – messages, as it were, from the cosmos. Of late, at the University of Iowa, Van Allen had been pioneering an instrument that he dubbed a 'rockoon', in which a balloon lifted a rocket up some 50,000 feet before the rocket fired and shot up another 200,000. By this means, in 1953 he had discovered negatively charged particles, electrons, in the sky and wondered whether they had something to do with making the auroras.

But the satellite was the most exciting scientific prospect he had ever encountered. He had designed an instrument that would ride as high as humans could imagine reaching; it would bear a simple Geiger counter, an instrument for measuring radioactivity.

Cosmic rays were simply a form of radioactivity, and Van Allen's Geiger counter would be able to fly through them, crackling as it encountered each one and telling him something about their history. And now it was to fly, under the guise of the army Jupiter C programme, which had been renamed *Explorer I*. That was perfect. Because *Sputnik* (Russian for 'satellite') had simply flown. *Explorer I*, however, would *explore*.

On the morning of Friday, 31 January at Cape Canaveral, the countdown began. Launch would be at 10:30 that evening. All went almost embarrassingly smoothly. At 9:45 P.M., somebody noticed a leak in the rocket's tail, but fixing it delayed the flight by only fifteen minutes or so. At 10:48 P.M., the Jupiter C rose on its haunches and launched itself into the sky.

There were four stages to this behemoth: the first, a liquid-fuelled rocket for propulsion; the second, a cluster of eleven motors; the third, three more motors; and the final stage, a single motor, and the home of the clinging payload. As they watched each stage ignite in its chosen succession, the engineers noticed that stage four looked a little fast. The satellite had certainly gone somewhere, but they didn't yet know where. True, it hadn't crashed back to the ground, but it might have been thrown like a slingshot to another ignominious patch of Earth. Until someone received its signal, no one would know if it had worked.

Van Allen had calculated when he should expect the news. One full orbit ought to take ninety minutes. The satellite should then be beeping over northern Mexico, where plenty of stations from Southern California would pick it up. If, that is, it was there. For the next hour he stood with the other nervous guests of the Pentagon War Room, which was now the satellite's centre for communications. A few amateur sightings trickled in, none of them especially credible. More coffee, more waiting. Another half-hour passed, and now even desultory attempts at conversation had ceased. The expressions were stricken, the mood was disappointed and dazed. Then the telephone rang. Almost two hours after launch, news had arrived from professional radio stations in Earthquake Valley, California, with these magical words: 'Goldstone has the bird.'

The room exploded with elation. Van Allen, Von Braun and the Jet Propulsion Laboratory's director, William Pickering, were immediately taken by army car to the National Academy of Sciences and smuggled in through the back door to make their report. Then came the press conference. Van Allen was amazed to find that the room was packed, even though it was now 1:30 in the morning. He later described the meeting as 'spirited'. Pictures of the three men, Van Allen, Pickering and Von Braun, brandishing a full-scale model of *Explorer I* above their heads, were soon beamed around the world. The two rocket men could scarcely contain their smiles. Van Allen looks quietly pleased, if perhaps a little tired.

In the days that followed, there was scarcely time to analyse the trickle of numbers coming in from *Explorer I.* There did seem something odd about them, though. Some of the time, the Geiger counter aboard was picking up the odd cosmic-ray blips, just the numbers you might expect. But once in a while the number dropped to zero, as if the machine were periodically malfunctioning. The trouble was, the method for beaming the numbers back wasn't working very well, and they couldn't get a continuous orbit.

*Explorer II* should have been better, but unfortunately she expired on the launch pad thanks to a faulty fourth stage on the rocket. But *Explorer III* was up in no time, on 26 March 1958, and confirmed what Van Allen was beginning to suspect. Either something was wrong with his instrument, or something was very strange about the sky.

Shortly after the launch of *Explorer III,* Van Allen flew to Washington DC. A receiving station in San Diego had downloaded an entire orbit of measurements from the satellite as it zipped overhead. Van Allen wanted the numbers. He picked up the tape from the Vanguard data centre on Pennsylvania

Avenue and took it back to his hotel. Until three in the morning he worked, making calculations with his slide rule and plotting the results on a piece of graph paper with ruler and pen.

As he studied the graph, Van Allen realised why the *Explorer I* data had been so erratic: they had been receiving read-outs from different bits of the cycle. But even though he now had the entire record laid out in front of him, it was still baffling. At low altitudes, the Geiger counter registered a modest fifteen to twenty hits per second, just what you'd anticipate for impinging cosmic rays – judging from his previous experiments with the highest-altitude rocket balloons. But then the instrument flat-lined. It was as if the higher up you got, the *fewer* cosmic rays you saw. That simply didn't make sense.

Van Allen packed up his results and went to bed. The next day, he went straight into the office and brandished his graph at two of his colleagues, Ernie Ray and Carl McIlwain. What, he wanted to know, did they make of that?

McIlwain, too, had been busy. He had spent the previous day testing the prototype Geiger counter, and he had momentous news. Of course the machine would flatline if it had no signal. But it would also flatline if it had *too much* signal. It would saturate at a count of 25,000 hits per second. The other two stared at him. That meant the intensity was ten thousand times what they had expected.

'My God,' said Ray. 'Space is radioactive.'

And not just any old space, but the stuff on the fringes of our own atmosphere, right above our heads. It was as if they had just discovered a mushroom cloud perpetually menacing us all from on high.

But if that was true, why don't we all fry? The explanation was already there, it turned out. A young scientist had proposed it in Norway more than sixty years earlier, though few outside the country had believed him.

*

**6 February 1903**
**The Festival Hall**
**Royal Frederik University, Kristiania, Norway**

The university's most ornate room was always an impressive sight: all Corinthian pillars, marble arches and polished wooden floors. But tonight, an unusually distinguished audience graced its long curving benches. Under the bright chandeliers, the cream of society in Kristiania (as Oslo was then known) was murmuring expectantly. There were industrialists, men from the worlds of banking, shipping and mining. The minister of defence was there. Indeed there was a particularly military flavour to the crowd, with the head of the Norwegian army also in attendance, not to mention various commanders-in-chief and the more martially minded members of Parliament; right at the front were representatives of Armstrong and Krupp, one of Europe's biggest weapons forgers. There was also a smattering of university professors and assorted Kristiania intelligentsia. Many of the men's wives had come, too, for this was a cold night in the middle of Norway's interminable winter, and to citizens tired of the musical recitals, theatrical plays and the odd séance that were the town's only other diversions, it promised to be an entertaining evening.

Kristian Birkeland was standing at the apex of the hall, the point towards which all the curved benches focused their attention. A slight, pleasant-looking man, he was wearing round wire-rim glasses, his ears perhaps a little larger than he would have liked, their size only accentuated by the thinning wisps of hair that clung to his temples. At least some hair remained there. Though he was still a young man, and much to his chagrin, the crown of his head had been bald as an egg for years. He was immaculately dressed, as always: long black frock-coat, waistcoat, snowy shirt, well-shined shoes and black bow-tie.

Birkeland waited for the crowd to settle down. He loved giving demonstrations and his talent for them, coupled with his evident ability in physics, had induced his Alma Mater, the Royal

Frederik University, to take him on as a lecturer. Now, at the age of thirty-six, he had been a full professor for five years, though appointees to this illustrious position usually had to wait until they were fifty or older.

That's one reason he had managed to persuade the university authorities to let him use their precious banqueting hall for what was really a private enterprise. For the assembled industrialists and military experts were here not to be impressed by developments in physics, but to see Birkeland's latest invention – and more importantly, to decide whether it merited their investment.

The demonstration would require a vast amount of electricity, which meant an equally vast generator. There was no room for this gargantuan machine in the hall, and besides, it didn't really look the part, so Birkeland had installed it outside in the university gardens. But its cables fed to the real star of the show: a brand-new 'electromagnetic cannon'.

The gun took centre stage. Its barrel had a bore of more than two inches, but the mysterious windings of copper coils that encircled it made it seem much larger. More impressive still, it was a full twelve feet long, robustly bolted to a large white frame. Inside the barrel, a hefty iron 'bullet' weighing some twenty pounds was poised ready to strike its target: a solid plank of wood five inches thick. Trial runs had gone well; a flick of the switch and the projectile had repeatedly zoomed out of the barrel and slammed into the bull's-eye.

The technology behind the gun was close to Birkeland's heart. He was fascinated by the burgeoning new science of electromagnetism. Earlier in his career he had studied in Paris with Henri Poincaré, one of the world's most famous scientists. And there, he had come across a set of extraordinary equations.

These were the same equations that had inspired Heinrich Hertz, and were even now electrifying Oliver Heaviside over in his rural English retreat. Back in 1873, when Birkeland was only six years old, Scottish scientist James Clerk Maxwell had produced a set of fundamental laws for how electricity and magnetism were entangled. He had put together all the discoveries about these two forces that had been trickling out over the previous decades. Electricity, it seemed, somehow affected magnets: if you hold a compass near a wire that is conducting electricity, the compass needle will lurch. The opposite also holds true: take a simple piece of copper wire and move it past a magnet and an electric current will immediately begin to flow through the wire, even when there's no battery or power source in sight. Electric fields somehow beget magnetic ones and vice versa, and this is what Maxwell's equations had captured.

The relationships embedded in Maxwell's equations also meant that electric and magnetic fields could reproduce each other in endless serpentine waves. That's what electromagnetic waves like light and radio waves were made of, all simply squeezed or stretched variants of the same interweaving fields.

As well as advances in physics, the equations presaged plenty of inventions: Marconi's telegraph, Bell's telephone, and also electric dynamos and motors. Because another aspect of Maxwell's equations showed that if you put a conducting object in a magnetic field, the object would move.

That's what had given Birkeland the idea for his cannon. What if, instead of using gunpowder, you could blast projectiles through the air with the power of electromagnetism? Surely that would be worth a fortune.

The money was what he was really after. Though Birkeland was entertained by his technological tinkerings, he was engaging in them to fund his true but expensive passion. He had become riveted by the question of what causes the northern lights.

Back in the 1890s, when he had begun his university research, Birkeland had been investigating a recently discovered phenomenon: cathode rays. These were invisible rays that streamed from a hot cathode into a vacuum, announcing their presence only when they hit the glass wall of the chamber, which had been coated with fluorescent paint and would glow a ghostly purple or green. (This is exactly the same principle used in televisions, before the flat-screen revolution. The normal bulky box of the 'tube' contains a hot cathode, which pours invisible rays through an evacuated chamber to paint an image when they hit the inside of the screen.)

When Birkeland started working on cathode rays, neither he nor anyone else knew exactly what they were. Like many of Maxwell and Hertz's electromagnetic waves, cathode rays were invisible and energetic. But in one crucial respect they were very different. Put a magnet near x-rays, or light, or wireless waves, and nothing happens. Their own in-built magnetic fields swamp anything the magnet can do, and the waves plough on undisturbed.

But cathode rays are different. When Birkeland put one of his magnets anywhere near them, the cathode rays slewed around and headed for the magnet's poles. This gave him an idea. He coated a magnetic object with fluorescent paint and sent a beam of cathode rays heading straight for it. As he had anticipated, the invisible rays suddenly appeared, dancing in glowing sheets of light over the magnet's north and south poles. They looked a bit like the aurora borealis, the 'northern lights'.

The phenomenon of sheets of ghostly green, red and white curtains billowing over the poles had been known about for centuries, and attempts at explaining these lights were as bizarre as they were numerous. But staring at his glowing magnet in the vacuum chamber of his lab, Birkeland became convinced that the truth was, if anything, even stranger. The sun was hurling beams of cathode rays in our direction. Earth's own magnetic field then captured these beams and directed them to the poles, where the air soaked them up and glowed with their energy.

A year after Birkeland came up with this idea, in 1897, British scientist J. J. Thomson discovered something else important about cathode rays. They weren't rays at all, or at least not steadily moving waves. Instead they were streams of tiny negatively charged particles – what we now know as electrons.

This would turn out to be most significant. Because if

Birkeland was right, the sun was flinging electrons and perhaps other positively charged particles towards Earth. These charged particles are a form of something we have learned to fear: the menacing radiation also created in a nuclear explosion. Though he didn't know it, Birkeland's suggestion was about much more than what creates the auroras. His hunch would lead to the discovery of exactly how our atmosphere protects us from the radioactivity that fills the rest of space.

Immediately, Birkeland had decided to set about trying to prove his hunch. However, what he had in mind would be very expensive: polar expeditions, new auroral stations, measurements, and a laboratory the like of which King Frederik University had never seen. Even though Birkeland was already receiving a large share of the university's research budget, he would need much, much more. So he decided to turn to his own inventiveness to supply the missing funds.

Birkeland enjoyed inventing things almost as much as he enjoyed figuring out physics. By the end of his life he would hold more than sixty patents for items as diverse as electric blankets, mechanical hearing aids and a technique for hardening whale oil to make solid margarine.[2] But of all Birkeland's inventions, he had the highest hopes for his electromagnetic cannon. Also in the Festival Hall was a certain Mr Gunnar Knudsen, an engineer and member of Parliament who was one of the five partners in the Birkeland Firearms Company. Birkeland had written to Knudsen two years earlier, inviting him to join the company he was forming:

> I have just recently invented a device with which it seems possible to use electricity instead of gunpowder as a propellant . . . Colonel Krag, who has witnessed my experiments, has proposed that a company be formed, consisting of a few men who will furnish capital to build a small gun according to my plans . . . Naturally, it would mean playing a lottery, but the contribution would be comparatively small, while I believe the chances are good for significant gains.

Knudsen knew and liked Birkeland, and had already supported some of his basic research. He replied good-naturedly: 'I accept with pleasure your invitation to participate in your invention, and promise, even if the big lottery does not appear, to keep smiling.' Under the circumstances, that was to be just as well.

It was almost time for the demonstration to begin. Birkeland was an expert showman. Although his job at the university was supposed to involve teaching as well as research, he rarely had time to spare these days for lecturing and he had taken to paying someone else to do it for him. But on the early occasions that he did grace the lecture hall the students had always loved it, largely because they were never sure what would happen next. His assistant, Olaf Devik, attended many of Birkeland's early lectures and recalled them vividly: 'He operated scarce electrical lecture equipment far beyond its rated capacity and burned out fuses with dignified nonchalance. Then he would stop in a royal manner, untie the ruffles of his ermine jacket and dry his glasses in order to better see his latest miscalculation on the blackboard.'[3] Birkeland wasn't above blowing fuses deliberately and for effect. Sometimes, he would reach over and almost caress a switch gently for a moment before suddenly pressing it to create a flash of light that made the audience gasp. Then, with the hint of a smile, he would straighten his ruffles and carry on with the lecture.

But for his electromagnetic cannon, the drama was to be in the silence. This was a device that could hurl a torpedo through the air with all the force of a modern weapon of war, and yet with the grace of a bow and arrow. There would be no explosion, no flash, no recoil; just as in the practice runs, the twenty-pound projectile was to emerge smoothly and silently from the gun's barrel, before heading with unerring precision towards its target.

Apart from the narrow safety corridor that Birkeland had railed off between the gun and its target, every seat in the house was filled. (Arctic explorer and professional daredevil Fridtjof Nansen insisted on sitting inside the safety area, and to

Birkeland's exasperation he flatly refused to budge.) Birkeland judged that the time was right to begin. 'Ladies and gentlemen,' he said, 'you may sit at rest. When I turn down the switch handle you will not hear anything but the bang of the projectile hitting the target.'

He reached for the handle. As he turned it downward, a deafening roar filled the hall. The flash was blinding; a flame tore out of the gun's barrel. The gun had short-circuited, sending a full ten thousand amps of current arcing across the metal casing. Poor Nansen's reaction, sitting as close as he was to the cannon, is sadly not recorded, but the rest of the audience panicked. There were screams of terror, followed by an undignified scramble of dignitaries struggling to escape the crowded hall. 'It was the most dramatic moment in my life,' Birkeland said later. 'With a single shot I dropped my shares' exchange from 300 down to zero.'[4] Unnoticed by the fleeing audience, the projectile did indeed hit the bull's eye, with a thud.

The next day, all of Kristiania was talking about the Festival Hall fiasco. Many of Birkeland's colleagues delicately avoided him. Some even gloated among themselves – it was about time this cocky young man was brought down a peg or two. A lesser man would have been dismayed, but Birkeland couldn't help but find the situation funny. After all, if you have to go down, it might as well be in flames. The question was, what to do next? The short circuit itself would be simple to fix, but the sensibilities of his potential investors might be a bit harder to patch up.

And then, before he could even make the attempt, he discovered a different use for his accidental spark. A week after his demonstration, at a dinner party hosted by Knudsen, Birkeland met industrialist Sam Eyde, who told him about nitrogen fertiliser. All plants need nitrogen, but if you want to grow them intensively you have to supply the stuff yourself. At the time, the only way to do this was to find natural deposits of 'saltpetre', a mineral-containing nitrate.

Whoever could artificially produce large-scale quantities of nitrogen fertiliser could revolutionise agriculture and potentially feed the world. Better still, there was a fantastic source of nitrogen just begging to be tapped, and as free as the air. Nitrogen makes up 80 per cent of our atmosphere; it is the great diluter, the inert gas that stops oxygen from burning up the world. But the trouble for Eyde lay in its very inertness. In the air, nitrogen exists as a molecule, its two atoms so tightly joined that almost nothing can separate them. From an agricultural point of view, as long as it stays trapped in this form, it is useless.

Eyde had the power to rip nitrogen molecules in two – he owned several of Norway's mighty waterfalls, which, via a hydroelectric plant, could create all the electricity he desired. But he had no idea how to turn his electricity into the sort of rapid, violent spark that he needed.

Birkeland, however, knew exactly what to do. He had terrified half of Kristiania with just such a spark. At the dinner party, he lit up with enthusiasm as he explained his idea to Eyde.[5] With his mighty sparks and Eyde's power source, the two of them could pluck fertiliser directly from the air.

Birkeland put his aurora research on hold. For the next three years he threw himself into the problem of turning his accidental short circuit into a fully functional furnace for splitting nitrogen. It was a brilliant success, and brought an enormous amount of attention from around the world. Cartoons showed Birkeland, in immaculate suit, bow-tie, glasses and curling moustache, solemnly turning a mangle that wrung dung from the sky, while bystanders held handkerchiefs to their noses and complained of the smell.[6] The money soon began pouring in. Now he could get back to his auroras, and spend it.

**12 January 1570**
**Bohemia**

First, a black cloud like a great mountain appeared where several stars had been shining. Above the cloud there was a bright strip of light as of burning sulphur and in the shape of a

> ship. From this arose many burning torches, almost like candles, and between these, two great pillars, one to the east and one to the north. Fire coursed down the pillars like drops of blood, and the town was illuminated as if it were on fire. The watchmen sounded the alarm and woke the inhabitants so they could witness this miraculous sign from God. All were dismayed and said that never within the memory of man had they seen or heard tell of such a sinister sight.[7]

Nobody who has seen an aurora will ever forget it. The light appears out of nowhere, usually a pale green colour, in shimmering curtains or jagged rays, or spirals that curl across the sky like the tracings of a giant snail shell. One of the eeriest things about auroras is that they are soundless. When you see them, you feel that lights like these in the sky should be accompanied by bangs; think of lightning, fireworks or bombs. But these lights are utterly silent, pulsing in and out of existence like the noiseless kneading of a cat's paws.[8]

Throughout the centuries since the lights were first recorded, they have inspired fear and awe in almost equal measure. Normally, they appear only in the far north or south, dancing over the polar snows during the long winter darkness. When the lights are at their brightest you can read by them, or make out faces inside an otherwise dark hut. They cast shadows. They light your way while hunting. Some say they were created by God for the people of the polar regions, as compensation for the annual loss of sunlight.

Legends abound. The lights flash from the swords of heavenly warriors, the shields of Valkyries, or the flailing wings of swans trapped in the ice. They are dead old maids, dancing and waving white mittens. (This is from western Norway, and a version of it persists today; it is still sometimes said of elderly spinsters that they will soon be off to the northern lights.) Some believe that waving a white cloth will make the lights stronger; others that waving or whistling at them will bring their wrath down upon you.

Many find them menacing. The lights will tear out your hair if you walk beneath them unveiled; they will take your children's heads and use them as balls to kick around the sky. They are a terrible portent, harbingers of war, poverty and plague.

This last fear arose most often when the lights escaped their usual bounds and made a rare foray to more southerly latitudes, where the people were unused to their flickering. Away from the poles, the white and green is often tinged with a violent red, as it was in the scene that terrified sixteenth-century Bohemia. These fears persisted even long after the superstitious Middle Ages had given way to more enlightened times. On 9 September 1898, red and orange auroral lights appeared without warning over the skies of London, Paris, Vienna and Rome, leading many to fear imminent disaster. The next morning, this portent seemed to have been confirmed, when news broke that the beautiful and well-loved Empress of Austria had been stabbed to death by an Italian anarchist.

Birkeland, of course, had no patience with such superstition. He sent an immediate telegram to an astronomer he knew, asking what was happening to sunspots around the same time. Soon, he received the answer he had been expecting. Just a few days before the auroral display, several unusually large groups of sunspots had appeared and lingered on the solar face.

Since Galileo had first peered through his telescope in the seventeenth century, the sun was known to bear the occasional unsightly speckle on its surface. (Identifying them was one of Galileo's stated crimes against the Church, for how could God's almighty creation be imperfect?) And by Birkeland's time it was beginning to be clear that the sun was prone to outbursts that might be related to these spots. On 1 September 1859, British scientist Sir Richard Carrington of the Kew Observatory became the first human to see such a flare-up in action. He was in the process of making one of his regular observations of sunspots. Since regarding the sun directly was now known to be dangerous to eyesight, something that Galileo learned only too late,

Carrington was prudently projecting the sun's disk on to a plate of glass coated with distemper of a pale straw colour. Nonetheless, the image was quite detailed, being some eleven inches across. He was carefully noting the positions of the spots when, from a group clustered towards the sun's northern latitudes, there came two patches of intense white light.

At first Carrington thought there must be a hole somewhere in his apparatus, but he quickly realised he was witnessing something much more important. 'I therefore noted down the time by the chronometer, and seeing the outburst to be very rapidly on the increase, and being somewhat flurried by the surprise, I hastily ran to call someone to witness the exhibition with me, and on returning within 60 seconds, was mortified to find that it was already much changed and enfeebled.' Carrington calculated that in the space of five minutes, the two light patches had travelled some 35,000 miles.[9]

The effect of this flare was soon to be felt on Earth. Eighteen hours after Carrington's observation, a magnetic storm disrupted telegraphs around the world and auroras were seen far outside their usual bounds, in Hawaii, Chile, Jamaica and Australia. This made perfect sense. Explorers had already noted that auroras seemed to make compass needles swing unexpectedly. Indeed, back in the eighteenth century, a Swedish scientist named Olaf Peter Hiorter had spent an entire year recording the position of a compass needle every hour to see how it deviated when the northern lights flickered overhead. He did take the time off for two short trips home, in August and at Christmas, but he still took an impressive – one might say excessive – 6,638 readings.[10]

Hiorter had intended his research as a warning to travellers in the northern lands that they should not trust their compasses when the auroras came. But Birkeland saw in it a deeper significance. If he was right, and the auroras were caused by electrons streaming in from the sun, then Earth's magnetic field would naturally be affected. As Birkeland knew well from his work in the entangled world of electromagnetism, where there were moving electrons there was electricity; and where there was electricity there would be fluctuations in magnetism.

Moreover, those changes would be strongest around the poles. The magnetic field that surrounds our planet looks a little like an apple cut in half: its lines of force emerge from the South Pole, bend over the equator and disappear back into the North Pole. Such 'closed' field lines form an almost impenetrable magnetic barrier, and few electrically charged particles from space can cross their invisible force field. However, the lines emerging most steeply from the South Pole do not connect with their counterparts in the North. Instead, both poles have a smattering of field lines that point directly up into space. These, Birkeland believed, provided the opening his cathode rays needed. The cathode rays would spiral down the open-ended field lines like beads on a chain until they hit the atmosphere. On the way, they would set the magnetic field jangling, and when they arrived they would be soaked up by the air, making it glow with the ghostly flickers of the auroras. The northern and southern lights were simply the outward signs of our sentinel atmosphere at work.

Thus, on 16 September, Birkeland published an article in *Verdens Gang*, a Norwegian newspaper, entitled 'Sunspots and auroras: a message from the Sun'. The auroras that had given rise to such fear across Europe were not some ghostly portents of disaster, he wrote. They were the outward signs of something streaming towards us from our own parent star.

Birkeland knew he had a great theory, but he still needed to prove it. With the money coming in from his inventions, he decided to build the biggest, most sophisticated lab the university had ever seen. Before long, the room was so crammed with equipment that only people working on the experiments were allowed to enter; students had to bellow their questions from the door. Electrical cables were draped everywhere; a massive power generator occupied a full third of the room, along with banks of rechargeable batteries, cameras and hand tools. Birkeland became famous for the bangs, the flashes and the odd smells that emanated from his new kingdom. A university committee was supposed to inspect every room at least once a year, but this was one lab they never dared to go near.

Even for Birkeland and his team, the lab had its dangers. They had all become used to receiving the occasional electric shock, and had taken to working with one hand in their pocket so that any large shock would travel straight down their sides and not across their hearts.

In the lab, Birkeland had grown more eccentric in his dress. He still sported his smart suits and immaculate cravats, but he now often completed the ensemble with an Egyptian fez made of felt and matching slippers in red leather with long, pointed toes. To the credulous, he claimed the fez was to protect his head from harmful electromagnetic radiation. To others he confided that it kept his bald head warm.

He worked obsessively. One of the first words that anyone describing Birkeland used was 'tireless'. He had suffered chronic insomnia since childhood, and his solution was often to continue working through the night. When a project gripped him he would stop for nothing, not even to eat. One of his assistants wrote about him: 'I never knew another man so engaged with science and with such reckless devotion. He worked far beyond the resources a human constitution can tolerate. It never occurred to him not to concentrate wholly on work.'[11] (This same attitude had cost him his marriage. During his work on the furnaces, Birkeland had married a teacher four years his senior, but she was unable to tolerate his work habits and left him not long afterwards. Birkeland wasn't overly concerned when she left – just anxious to make sure that everyone knew it was his fault, and that she had plenty of money. He might have overdone it, since though she neither remarried nor took another job, she spent the rest of her summers sunning herself on the French Riviera.)

And yet for all his intense focus, Birkeland could be maddeningly abstracted. Sometimes in the lab he became unexpectedly distant; he would leave his assistants working and wander out into town for an hour or two without explanation. When he returned, his hat would be pushed to the back of his head and he would burst in full of excitement about some new idea or insight.

Birkeland also had a fundamental disdain for bureaucracy. He kept neither diary nor notebooks, but simply scribbled notes on scraps of paper that he stuffed in his pocket or lost under cushions or left to float around the lab. Though he fortunately had an excellent memory, he still must have tried the university administrators beyond endurance. When asked to document the details of his expenses, he replied, 'What for? I remember the sum.' He would send the officials small notes declaring that he had taken over this or that room for a new laboratory. Once he even appropriated half a lecture hall. 'By putting the students closer together there is enough space in the reduced hall' was the casual way he informed the infuriated dean and vice-chancellor. He assuaged their fury by paying for all the alterations himself.

The most impressive thing about Birkeland's new lab was the vacuum chamber that held his artificial sun and earth. It was a full cubic metre in volume, its sides straight like those of a glass aquarium but with curved corners and walls two inches thick to prevent it from collapsing when the air was sucked out. Indeed the chamber was big enough for his most slender assistant to climb inside and sit cross-legged to clean the walls. (Birkeland once teased him by 'accidentally' trapping him inside. He loved practical jokes, and his assistants tolerated them good-humouredly. Once he set up an iron bar so it was highly magnetised, laid it on a metal table and then nonchalantly asked one of his assistants to move it. After the assistant had struggled for a few moments, the others joined in and, by heaving together, managed to shift the bar an inch or so. They were so involved in their task that nobody noticed Birkeland as he flicked off the switch that was powering the magnet, and bar, assistants and all went flying off the table.)[12]

Inside the chamber, near one of the walls, a glowing cathode generated beams of electrons to imitate the sun. The rays then barrelled invisibly through space until they encountered Birkeland's terrella, which is Latin for 'little earth'. This was a thin brass sphere, about fourteen inches across, inside of which was an electromagnet made up of an iron core wound about with copper wire. For extra authenticity, Birkeland had even ensured that the magnet was tilted by 23.5 degrees, so it would match Earth's own tilt.

The surface of the sphere was coated with barium platinocide, a chemical that glows when hit by electrons, just as television screens do today. When Birkeland ramped up the terrella's internal magnet, electrons from the cathode 'sun' would veer off towards the terrella's poles, where they would dance in two rings, one north and one south, with a ghostly purple glow.

Birkeland was delighted with the spectacle. He wrote: 'It will be easily understood that in addition to the purely scientific reasons for doing this, I have also a secondary object, which is to give myself the pleasure of seeing all these important experiments in the most brilliant form that is possible for me to give.'[13]

From time to time Birkeland squeezed an audience into his lab to show off his artificial auroras, and few failed to be amazed. Here was a man who could make the northern and southern lights dance at will. What's more, he could explain why they appeared. And his experiments matched his theory with a faithfulness that was thrilling.

Still, just because the lights dancing over Birkeland's terrellas looked like the auroras, that didn't necessarily mean he was right. To many, the notion of radioactive beams from space was absurd. Birkeland had to prove that what happened inside his vacuum chamber was also taking place out in the real world.

He needed to show that when spots appeared on the sun, electric currents appeared in the northern skies along with the auroras. He couldn't measure the currents directly, since they would be too high up. But he could perhaps detect their influence on the magnetic field around them. So, Birkeland packed up his instruments and went north.

For someone who was slight and relatively frail, Birkeland approached his polar expeditions with a zest that was almost reckless. Of course, he needed to be in the auroral zone, which meant the farthest north of Norway's provinces. Also, he had to go in winter – which, with its long, dark nights, was the best time to study auroras. But to make matters even harder, he had also decided to study his auroras from the top of a high mountain.

One reason was to be as close to the lights as humanly possible. Indeed, some theories held that auroras came from electricity that leaked out of Earth via the point of a mountain, like an inverse lightning rod; even if, like Birkeland, you didn't believe this, there were many who thought that in northern Norway auroras came down so low that they touched the mountaintops themselves. True, this was because of rumours and legends rather than scientific data, though there was also the occasional eyewitness report. Some of the 'close encounters' were described with amazing detail and poetry. This is from a boat trip at Talvik in Finnmark, as late as 1881:

> Immediately after nightfall the northern lights began to flame merrily in the sky. They gathered like a huge fire in the deep-blue vault of the heaven, and great pencils of rays, mauve, blue and green, united and danced in flaming witches' knots above the boat. We had just about reached the middle of Korsfjord when I suddenly noticed an aurora above Alta that had knotted itself right down on the surface of the water and was rushing across the fjord at high speed . . . 'It will overturn the boat,' shouted Jakob from the thwarts. And in the darkness I could see the men bend their backs and heave on the oars so that the phosphorescence shone about the blades . . . I shut my eyes for an instant. When I looked round a moment later with the light penetrating my eyelids, I found we were in the midst of such a fantastic sea of light as I shall

> never forget. Flames of wonderful transparent colour surrounded us, violet, blue and green, but without the slightest breath of wind . . . That rare but lovely play of light passed us by in a few seconds. A moment later it was gone.[14]

Even if – as Birkeland suspected – these reports were spurious and the aurora didn't touch the ground, the chance of being engulfed by an aurora was irresistible, and a mountaintop seemed the surest way of trying.[15] In February 1897, Birkeland headed off with two assistants to Finnmark, in the far north of Norway, on a trip to seek out suitable mountain sites.

At first, all went well. There was scarcely any daylight so far north at this time of year, but the moon shone brilliantly on the mountains' thick blanket of snow, lighting the way for the strings of reindeer that pulled Birkeland's team and equipment up the slopes. On 9 February the wind picked up a little, blowing the snow crystals along the ground like trails of smoke. But it was nothing much to be alarmed about, even if the temperature was a chilly –25 degrees Celsius. There were only another sixteen kilometres to go to reach their destination, Lodikken Hut.

However, as they pressed on with their Finnmark guide the wind grew stronger, always in their faces, blowing, it seemed, directly from the hut they were trying to reach. Birkeland began to be alarmed. The small team struggled on. They were no longer splendidly skimming over the snow in their sleighs. Their guide was now forced to lead the reluctant reindeer, and anyone foolish enough to sit in the sleigh would have been pelted with stones and shards of ice that had been whipped up by the wind.

Eventually the reindeer lay down and refused to go farther, and the guide, his face now white with frostbite, climbed under his Finnish furs and similarly refused to budge. Nothing remained but to use the baggage and sleighs to build a barricade against the wind and then pitch a small tent behind it. This was supposed to be a trip of only a few hours, so aside from the emergency tent the team had little in the way of supplies – no food or fuel to melt water, even if they could have lit a stove in this appalling wind. For the next twenty hours, throughout the long, dismal polar night, the small team huddled in their sleeping bags, trying to ignore their gnawing stomachs and keep their tent from being buried by snow. When dawn finally came, though the wind had scarcely abated and Birkeland could still not see more than a few yards ahead of him, he decided their only chance for survival was to try to find a route back down the mountain.

Gradually, grudgingly, the guide stirred to help break the miniature camp and set up the reindeer. But then, when he had thawed out a little, he proved his worth. Birkeland recalled, 'The couple of hours spent in the descent were the most exciting I have ever gone through. It was now that our guide showed himself to be the adept that I had been told he was. It was wonderful to see the way he ran to the right or to the left, to find tracks or take a course, and how he drilled the reindeer when they became unmanageable and suddenly set off up in the face of the wind again.'[16]

Thanks to this wild ride, the team arrived safely back at Gargia, thirty-one hours after they had left. It was a miracle that nobody had died. However, one of Birkeland's young assistants, twenty-year-old Bjørn Helland-Hansen, was suffering badly from frostbite. His hands were white and stiff from the tips of his fingers all the way to his wrists. While the others soaked up the warmth and comfort of the mountain hut, poor Helland-Hansen had to sit with his hands in iced water, waiting for the circulation to return – and with it the inevitable burning agony. He later lost the top joints of most of his fingers, and with them, his dreams of becoming a surgeon.[17]

But one good thing came out of this disastrous expedition. At ten minutes to six on 5 February, Birkeland had seen something that left him spellbound. The weather was clear, the moonlight shining brightly on the snow, and then, suddenly, an even brighter light appeared in an arc that curved across the sky from east to west. At first it was narrow and intense but then it draped into shimmering curtains, and bundles like sheaves of corn. Birkeland watched, awestruck, as the heavenly light-show played itself out for more than an hour.

And then, the next night, the same thing happened. At just after six, the aurora returned, and passed through exactly the same set of arcs, curtains and sheaves. To Birkeland this was almost an omen. Earth's daily twirl around its axis brings it back repeatedly to face the sun. So if the aurora could appear so consistently two days in a row, at exactly the same time, it surely had to be caused by something coming from the sun. His hunch from 1896 had to be right.

In autumn that same year, Birkeland returned to the mountains of Finnmark, determined to try again. This time, he found exactly what he was looking for in the district of Haldde, on the west side of the Alten Fjord. Atop two adjacent mountains Birkeland set about building stone and cement huts, the world's first permanent observatories for studying the auroras. He was inordinately proud of them:

> In clear weather everything that takes place in the sky can be observed, from the point where it begins to that where it leaves off. The view is uninterrupted, and from both observatories, but especially the highest and northernmost, there is a panorama stretching from the sharp, blue peaks of the Kvaenang mountains in the west, to the softer outlines of the Porsanger mountains in the east, and from the precipitous cliffs of Lang Fjord and Stjerne Island in the north to the mountain plateau in the south, stretching inland in undulating lines as far as the eye can see, in towards the winter home of the mountain Lapps. And far below lies the fjord like a dark channel.[18]

By September 1899 the huts were finally ready. Birkeland took two teams up the mountain, aiming to spend the entire winter. They would be there until April 1900, passing the first day of the new century in conditions that were sometimes benign, more often appalling. The winds were phenomenal, often raging at more than 100 miles per hour. 'It sometimes roared so against the houses, that you would have thought you were sitting at the foot of a waterfall; and the floors trembled and everything shook. We soon got to be able to gauge relatively the storm outside by the noise within.'[19] When the storms rose, nobody could leave the hut for days; if anyone did try, it took all three men, with great effort, to close the door. Inside the hut, even with the door closed, water would sometimes freeze a few yards from the glowing stove, and a lamp was once blown out while sitting on the table in the middle of the closed room. 'No one,' said Birkeland, 'who has not tried it can imagine what it is to be out in such weather.'

One person who repeatedly found himself caught out by these storms was the sturdy Finnmark postman who arrived once or twice a week with news of the outside world. 'We were often afraid for him,' said Birkeland, 'but he was always all right, though sometimes so covered with ice when he arrived that he was quite unrecognisable. I once asked him if he were never frightened when the weather was so bad. At first he did not answer, but sat quietly down to thaw; but a little while after he said: 'I'm too stupid to be frightened".'[20]

In spite of the storms, the expedition was a huge success. The team had seen aurora after spectacular aurora, and Birkeland's most precious instruments – his magnetometers – had exceeded all expectations.

The magnetometers had a room of their own in the stone hut. Before Birkeland entered, he would empty his pockets of coins, pocketknives or anything else that might disturb the delicate magnet at the heart of the instruments. Even the buttons on his clothes were made from bone, and his round glasses were rimmed with non-magnetic gold. All magnetic metals had been similarly banished from the room. The door hinges were made of brass, the nails were of copper instead of iron, and the heating pipes were ceramic.[21]

Each of the three magnetometers was continuously monitoring a different aspect of Earth's magnetic field. One recorded which direction the field pointed in, another the horizontal strength, and the third the vertical. They were housed in large boxes, each with a hole cut in the side. Inside each, a magnet dangled on a fine quartz thread, which also bore a mirror. Lenses focused the light from an oil lamp into a fine beam that entered the box, reflected off the mirror, and then exited to strike a roll of photographic paper. If the magnetic field overhead changed in any way, the magnet would swing in response, taking the mirror with it, and the reflected light would be deflected. Even if nobody happened to be in the room when this took place, any deflections would show up clearly as soon as the photographic paper was developed.

Throughout their long winter, almost every time the teams spotted an aurora overhead the mirrors on the magnetometers swung, and the line tracing their path lurched. Just as Birkeland had anticipated, he was marking the overhead passage of electric currents, the cathode rays streaming in from the sun.

It also became clear as the winter went on that the lights came nowhere near the surface, even of such lofty mountaintops. In one way, that was a shame; the bright-eyed Birkeland would have enjoyed the experience of being immersed in his precious lights. But at least it meant that his next expeditions could be closer to sea level.

For Birkeland already knew that there would be many more expeditions. The movements of the currents were obviously complex. To trace their sinuous pathways, Birkeland realised he would need measurements from sites that were much farther afield. So he decided to extend his operations. As well as in Norway, he would set up observatories in Iceland, Russia and the frigid northern archipelago of Svalbard. Meanwhile, data began flooding in from observatories around the world, a response to a circular Birkeland had sent out asking for magnetograms to be sent from any available station on the globe.

And as he mapped out his measurements and pieced together the jigsaw puzzle, everything Birkeland saw confirmed his suspicions. Sunspots coincided with beams of electrons that shot out of the sun like searchlights. Some missed Earth; most glanced off the closed magnetic field lines that curved over the bulk of the globe's surface. The rest were directed safely to the poles by our protective magnetic field. He couldn't see the electrons themselves, but he could trace their jangling influence as they spiralled down the magnetic field lines over the North Pole. Then they hit a short circuit. (Though Birkeland didn't know it at the time, this was the ionosphere, which as well as fielding incoming x-rays provided a handy horizontal conduit for the incoming electrons.) As the electrons passed horizontally above in great arcing currents, they were gradually absorbed by atoms and molecules of air, and yes, the northern lights glowed. Finally, any electrons that remained finished the loop by attaching themselves to new field lines, and spiralling harmlessly back up them and out into space.

Over the next decade, Birkeland collected more and more data in support of his theory. He published it in a grand opus, two handsomely bound volumes that made him, if anything, even more famous in his native land. He wrote: 'The knowledge gained, since 1896, in radioactivity had favoured the view to which I gave expression in that year, namely, that magnetic disturbances on earth and aurora borealis are due to corpuscular rays emitted by the sun.' And: 'It must . . . be considered as certain that the cosmic rays which come vertically towards the earth in such a way as to form auroral rays, are entirely absorbed by the atmosphere.'[22]

This should have been the prime of Birkeland's life and the apex of his success. But as he progressed through his forties, his earlier youthful vigour and optimism somehow turned sour. Since his early days in Paris he had occasionally been subject to depression, 'nervous freezing attacks' that would send him to his bed for days on end. He was now experiencing such attacks increasingly often; with them came paranoia, despair and ever-faltering health. He was also frustrated that his theory was not getting the international recognition he was sure it deserved. Part of the problem was that he had written in French. Not only was he a fluent speaker, but French had long been Europe's premier language of culture and natural philosophy. But this was the turn of the twentieth century, when the British Empire was at its peak and English was becoming the new lingua franca.

Moreover, English scientists were primed to look askance at Birkeland's ideas. Back in 1892 the great Lord Kelvin, one of Britain's most famous and distinguished physicists, had pronounced: 'There is absolutely conclusive evidence against the supposition that terrestrial magnetic storms are due to magnetic actions of the sun; or to any kind of dynamical action taking place within the sun . . . the supposed connection between magnetic storms and sunspots is unreal and the seeming argument between the periods has been mere coincidence.'[23]

Kelvin's views held great sway, because he was almost invariably right. Except, as it turned out, in this case. Still, British scientists stuck to Kelvin's guns. Birkeland, always an insomniac, was now having more and more trouble sleeping. To find any rest at all he became increasingly reliant on veronal, a sleeping agent that had the particularly important attribute of not adding to his mounting struggles with depression.

He went to Egypt and became convinced that he was being tracked by mysterious and sinister foreign agents. He decided to return home and cabled that he would be back in Norway in time for his fiftieth birthday, on 13 December 1917. The First World War was now under way, and he had to take a circuitous route via Tokyo. On the morning of 16 June 1917, one of the Japanese physicists Birkeland had stopped to visit found him dead in his Tokyo hotel room. He had taken ten grams of veronal, instead of the prescribed half-gram, and his heart had failed. It was perhaps an accident.

In his lifetime, Birkeland had received four nominations for the Nobel prize in chemistry and four in physics. An illustrious committee of Norwegian scientists was putting together what they believed was the strongest nomination yet for the physics prize when news reached Norway of his death. The project was quietly shelved.

Because of the war, nobody could come from Norway for the funeral. Instead, Birkeland's Japanese hosts presided over a Christian ceremony and cremation. At the service, one said, 'What Birkeland has achieved in the fifty years of his life is as brilliant as the dazzling waves of the aurora, which have exerted such a mighty attraction on him.'[24]

For decades after Birkeland's death, his theory remained in limbo. Even when the ionosphere was discovered and it should have been obvious that this was the conduit for Birkeland's currents to sweep over the sky, few scientists accepted his argument. Only in the 1960s was he finally vindicated. For this was now the space age, the time when satellites could penetrate the world that Birkeland had simulated and monitored, but could never touch. Satellites had discovered that space was radioactive. And they were also about to discover just how right Birkeland had always been.

**1 May 1958**

James van Allen was presenting his findings to the world. The patch of space hugging the top of our atmosphere was mysteriously radioactive. That's what *Explorers I* and *III* had clearly shown. He still wasn't exactly sure what this meant, or why it would prove to be important, but that hadn't prevented him from laying out the results to assembled scientists at the National Academy of Sciences. More difficult was explaining them to the journalists at the press conference that followed. Van Allen struggled to find the words. The radiation they had discovered seemed to congregate in a giant cloud, shaped like a doughnut with Earth occupying the hole in the middle. It was corpuscular radiation – that is, charged particles – girdling the planet in a giant, well . . . something like . . . 'Do you mean like a belt?' one reporter demanded. 'Yes, like a belt,' Van Allen replied. And thus the 'Van Allen belt' was born.

But there was still so much more to know. Where exactly did the radiation come from? How did it get trapped? What stopped it from continuing straight on down to Earth's surface? Van Allen already knew what he would do next to try to understand this. For as he hurried back to Iowa, he took with him a secret. Back in the spring he had received an extraordinary and highly confidential request. The army had decided to detonate nuclear bombs high in the air. Obviously it was vital that nobody knew. Their ostensible purpose was to see what might happen if, for example, the Russians did it first. They had frighteningly little idea of where the radiation from the bombs might go. But Van Allen's stock had sky-rocketed along with his *Explorer* satellites. Would he help? Could he perhaps design a satellite that would monitor the radiation, and in the process learn a little about how his newfound belt worked?

Yes, of course he could. Van Allen and his team immediately began work on *Explorer IV.* They had only a few months before the tests were due, and the pressure was high. But the army superiors believed Van Allen's opinions on space instrumentation to be infallible. A string of engineers and officials were obliged to trek out to his small lab in the middle of Iowa throughout the summer to hear his opinions on their plans. He was amazingly pleasant, one remembered: 'My most vivid memory of that visit was of a phone call he received from some important general while I was in his office. As I recall, his exact words were: "Yes, General, I would be happy to come to Washington to testify for your project next week. However, one of my students is taking his oral exams then and I have to be here to help him." From then on I looked on Van Allen as a voice of reason in a world gone mad.'[25]

Van Allen himself was calm about the whole affair: 'Visitors to the University of Iowa . . . were astonished to find that a crucial part of this massive undertaking had been entrusted to two graduate students and two part-time professors, working in a small, crowded basement laboratory of the 1909 Physics Building. But we knew our business, and were in no way intimidated.'[26]

On 1 August 1958, a ten-megaton bomb code-named Teak was exploded seventy-five kilometres above Johnston Atoll in the Central Pacific. Twelve days later came another, code-named Orange, and then three more at still higher altitudes. *Explorer IV* was there in the sky to see them all. She was the best satellite yet. A picture exists of Van Allen with her before launch, only the thinning crown of his head visible as he kisses her goodbye.

*Explorer IV* saw exactly what Van Allen had hoped for. Though the lower-altitude explosions disappeared without trace, the high ones formed a new radiation belt, above the first. This had to be how the belts formed – by incoming radiation being trapped by the field lines, just as he had thought. This new belt was faint and feeble, though, and lasted only a few weeks before it drained away. It must have been humbling to see how little mankind's mightiest weapon could do, compared with the natural forces that were already somehow subjecting our planet to their relentless attack.

The Americans detonated a few more high-altitude nuclear bombs; so did the Russians. Mercifully all such explosions were banned by the treaty of 1967. But Van Allen now had something else to occupy him. On 6 December 1958, a new spacecraft, *Pioneer 3*, left Cape Canaveral, bound for the moon. On board was another of Van Allen's Geiger counters. For the newly born NASA, the mission was a bust. At about 63,000 miles above the surface, *Pioneer 3* turned through a graceful arc and then tumbled back to Earth. But Van Allen was happy. The satellite had still gone farther than any man-made spacecraft before it, and in the process it had made another important discovery. There was not one Van Allen belt, but two.

The second, outer, belt now had all Van Allen's attention. It was much higher – some 10,000 miles above Earth's surface, while the inner one was a mere 4,000 miles up. It was also bigger, and the particles it contained were much more energetic. What had supplied the space above Earth with these two thick clouds of radioactivity? Had they come to us, as Birkeland had suspected, from the sun?

Today, fifty years after *Sputnik*, beeping satellites are almost ubiquitous in our skies. Some are for communications, some for military purposes, but many were put there, like *Explorer I*, to tell us more about the most tenuous edges of Earth's atmosphere. The space above our planet has turned out to be far more complicated and strange than even Birkeland, with all his vision, had guessed. But it is astonishing how much he divined, from a vantage point fixed so far below.

He was right that cathode rays, or rather electrons, do indeed come from the sun. In fact they appear in a continuous stream, which we now call the solar wind. The electrons aren't alone; they couldn't be. Negative charges repel one another, as do positive ones – only opposites attract. So a cloud of negatively charged electrons coming from the sun would spread out and disperse long before it could reach Earth. Instead the solar wind contains a mix of positive and negative particles – a plasma – which is flung off the wispy outer reaches of the sun's atmosphere at a temperature of a million degrees. This solar wind blows constantly from the sun, in every direction. It breathes on comets to draw their long tails out behind them. It crashes continually against Earth's magnetic field, like a stream flowing past a rock, squeezing the field lines in the front and drawing the ones behind into a long tail that continues for hundreds of thousands of miles beyond Earth's backside.

But the serious emissions, the ones that cause Van Allen's belts and Birkeland's auroras, come from something more tempestuous still. Once in a while the sun hurls out a monstrous glob of plasma called a coronal mass ejection. Birkeland imagined the event, but he probably had no notion of its scale. A single outburst can easily contain a billion tons of glowing hot plasma moving at five times the speed of the solar wind. Nobody knows why the sun does this, though – also as Birkeland suspected – it is somehow associated with sunspots. What's certain is that with frightening regularity these deadly clouds come powering towards us, surfing on the shock waves of the solar wind.

The first place to feel one of these solar torpedoes is the outermost force field of our planet's arching magnetism. Out front, the lines of force press inward under the strain, but they do not buckle. Frustrated plasma streams around the planet's sides and then back, pressing up against the long magnetic tail that stretches far beyond the dark side of Earth. On it urges and squeezes until some plasma manages to barge past the magnetic sentinel and arrive inside the tail. By now, the plasma has long overshot Earth itself. But the field lines in the tail are now stretched almost beyond endurance; they snap like an elastic band, catapulting the plasma back towards us.

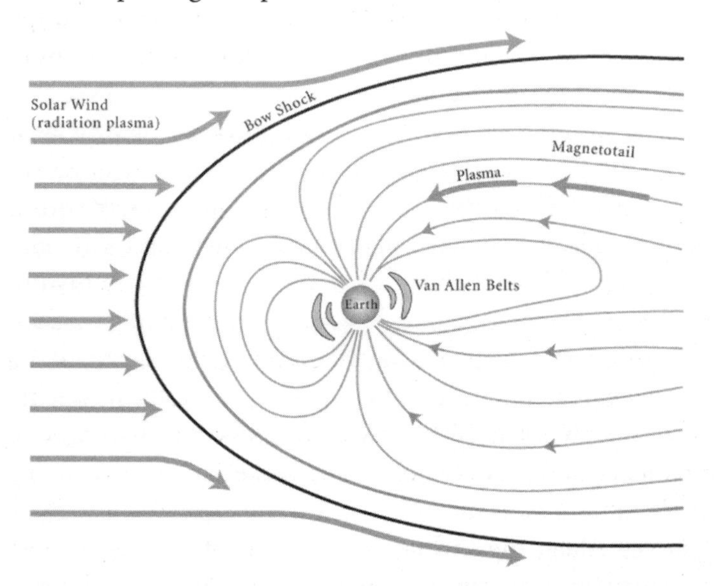


Solar radiation plasma finds its way around Earth's magnetic field and into the long magnetotail but is then directed to the planet's polar regions, where it is soaked up by the air or splashes out to create the Van Allen belts.

What happens next is a marvel. As fast as the plasma barrels back in towards Earth, more field lines gather it up and send its charged particles spiralling towards the poles like beads on a wire. And – just as Birkeland imagined – the electrons are then swallowed by the air of the ionosphere, which lights up with the effort to create the flickering lights of the auroras.[27]

Small wonder people have feared the auroras for millennia; they are the outward sign of horrifying attacks from space. But those people who revered them are right, too, for they also show that our protective air is doing its job well.[28]

The Van Allen belts are an intricate part of this system. At first, Van Allen himself thought they were a 'leaky bucket' that caught the plasma and held it until the bucket overflowed. We now know they are more of a splash screen. Plasma that is too energetic to be channelled to the poles and dealt with by the air instead bounces up into the outermost Van Allen belt. The field lines that arch some 10,000 miles above Earth hold these particles in suspended animation; they are unable to escape back into space or menace the ground, but leak harmlessly away and are replaced by new ones.[29]

Birkeland would have been proud to know how right he was. He also would have enjoyed the pride of place he now holds on the Norwegian 200-krøner note. The front shows him with a typical half-smile, wearing his smart suit and his round, wire-rim glasses, though sadly not the red fez. On the left is a miniature sketch of a terrella, and behind him a stylised aurora. The back shows a geographic map of the Arctic, marking overhead the locations where satellites discovered flowing electrons in the sky. These are exactly what Birkeland predicted with his magnetic measurements, and they are now called Birkeland currents in his honour.[30]

James van Allen, meanwhile, became one of America's most famous scientists, appearing – among other illustrious places – on the cover of *Time* magazine. He also, of course, left his own name written in the sky in the radiation clouds that float over our heads. But he did even more than that. In 1973, while working in the clean-room of a satellite called *Pioneer 10*, Van Allen surreptitiously whipped off his white glove and planted a fingerprint on the spacecraft.[31] *Pioneer 10* was the first man-made satellite to encounter Jupiter, and then Saturn. It continued on, to the outer reaches of the solar system, and beyond. By Van Allen's ninetieth birthday in 2004, *Pioneer 10* had travelled over 100 billion kilometres. He died, aged ninety-one, in August 2006, but the satellite continues to coast silently into the deepest of deep space, heading for the red star Aldebaran, which forms the eye of Taurus, the Bull. The journey will take more than 2 million years, and Van Allen's fingerprint will go, too.

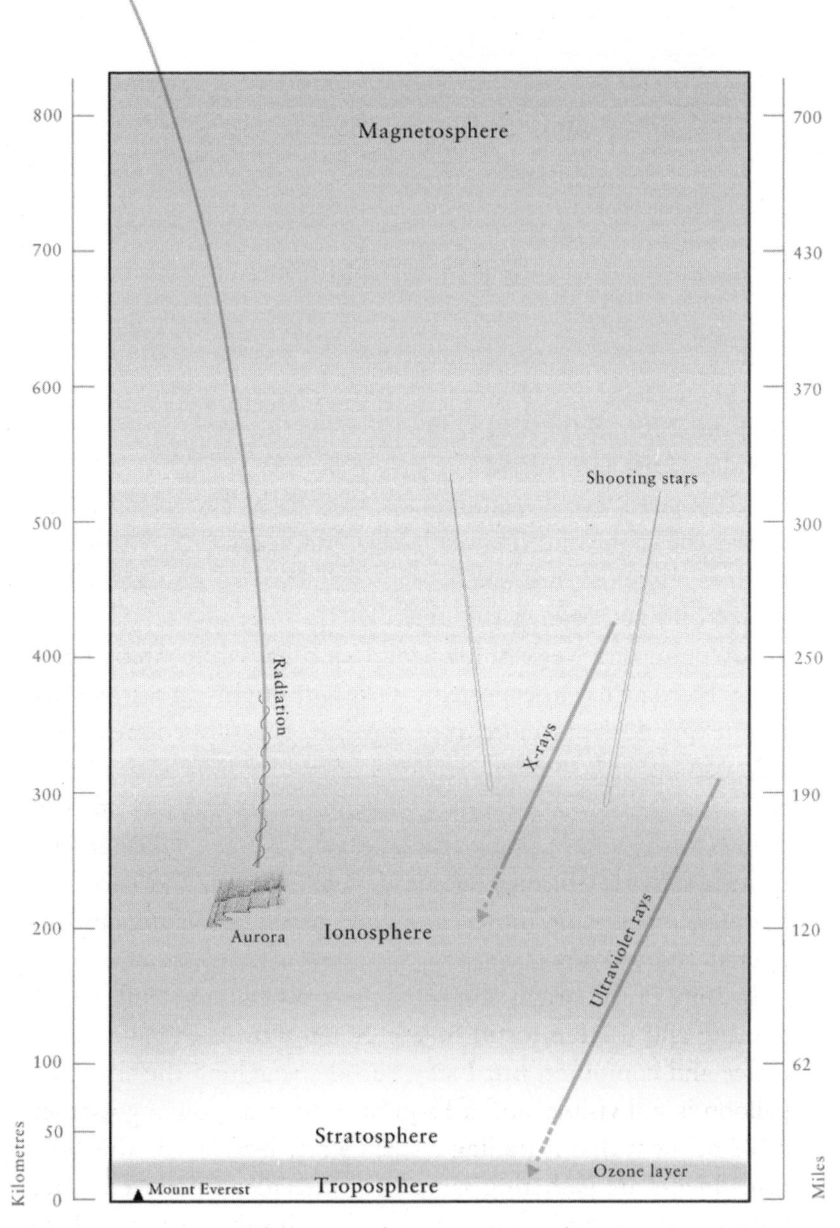


The atmosphere to true scale.

# EPILOGUE

**16 June 2006, 9:00 A.M.**
**Tasiilaq, East Greenland**

'There it goes.' Local weatherman Søren Basbøll opens his fist to release the cable, and his weather balloon leaps eagerly upward. Within a few seconds we have to crane to see its tight white sphere outlined against the blue.

After almost a week of low clouds and flat white Arctic light, today has dawned unexpectedly clear and bright. At last we can see the tops of the mountains on the other side of the fjord. They are lined up with serrated sides and pointed summits, the way a child might draw a mountain range, and their flanks are still draped with snow. I know that beyond this first rank there are countless more, nothing but mountains, glaciers and snow for 1,000 kilometres or more, and most of the peaks are unconquered and unknown.

In spite of the bright sunshine, the temperature is still below freezing and there is a stiff breeze. Søren retreats indoors to his coffee and computer, but I stay outside, watching the sky. The balloon is still visible, and if I squint, I can make out a gossamer cable below it that is trailing a small white box. Inside this box, instruments are gathering samples of Greenland's air. They are tasting and testing it. The numbers are already surfing back to ground on Marconi's radio waves: temperature, pressure, wind speed and moisture, the meat and drink of the weather world.

I check my watch. By now Søren's balloon will have passed through the lowest part of our atmosphere, the part that means the most to the local Inuit hunters, and to the rest of the human race.

The air here in Greenland seems somehow more substantial than it ever is farther south. Perhaps it's the cold, or the manifest absence of smoke and industrial pollution, but you feel aware with every breath that something sweet and fresh is entering your mouth and filling your lungs.

Robert Boyle would have had little difficulty in persuading the residents of this small Arctic town about the power of the atmosphere or its substance. They feel it almost every winter, when the Piteraq winds appear. Heavy, cold air pours down from the icecap; it gathers speed as it is funnelled down the glacial valleys until it arrives in town with the force of a hurricane, to tear off roofs and shatter windows. During a Piteraq is the only time the sledge dogs are allowed to roam free from their chains and seek shelter where they can – the air can be more than strong enough to hurl them into the sky.

Søren is Danish, but he has lived here for decades, and he has developed his own private scale for the Piteraq winds. Level one is when his oldest anemometer – a type of wind-measuring device – gives up the ghost. Level two, at about one hundred knots, is when the second one follows suit. Level three is when the welding of this second anemometer breaks loose and begins to flap uncontrollably, like a helicopter with only one rotor. And level four is when all his instruments blow away.

We are approaching summer now, and the winds are gentler. Still, today's breeze was enough to transform the bay. Yesterday the water was practically clear; today it is crammed with floes of ice that have been blown in from the open sea. Years of partial melting and refreezing, slamming together and ripping apart, have sculpted these floes into the shape of whipped meringues. And here and there, an iceberg towers above the sea ice. Each one of these tall bergs was born in the heart of this frozen land and has spent thousands of years sliding gently towards the coast before it picked up pace on the steep side-slopes, broke off into the water of the fjord and is now sailing out to sea.

Greenland is a sensitive soul, a highly strung region of a restless planet. A slight nudge downwards in temperature can make the entire fjord freeze; another nudge and there's nothing but open water. I'm here to witness some of these changes, effected by the atmosphere. For Greenland's sensibility means that the sinister side of carbon dioxide is already making itself felt. The icecap here is melting; the glaciers are speeding up and delivering ever more icebergs to the sea; the sea ice, haven for seals, hunting ground for polar bears, is on its own downward route to extinction. Satellites show that the summer minimum in sea ice has shrunk by 8 per cent per decade over the past thirty years, and by the end of the century summers here will be all but ice-free.

The Inuits here say they are unconcerned by this prospect. They are hunters, not farmers. They are accustomed to change, to reading the signs of our restless planet and reacting to them. Theirs is an entrepreneurial attitude to weather and climate, one that the rest of us will do well to learn as the changes affecting Greenland make their inevitable way south.

But Greenland also counts on the natural warming that carbon dioxide brings and the capacity of the air to redistribute the world's warming wealth. William Ferrel's third giant wind cell descends hereabouts, bringing with it warmth from the south. There's not much life on view on this frozen island, but without Ferrel's wind cells there would be no life at all.

Above the layer of wind and weather, Søren's balloon must now be swelling as it rises into the thinning air. On the ground it was only about a metre wide, but it will eventually grow to four times the size, as the hydrogen inside meets less and less resistance when shoving against the thin latex skin. Perhaps it has already passed through Joseph Kittinger's vantage point, twenty miles up. Around it is the blackness of space; in the distance, the gentle curve of Earth with its thin blue line of atmosphere; below it, a few wispy clouds, and me.

It must be encountering our planet's outermost defences. There might even be an aurora, flickering into life in the black sky above Søren's balloon. On the ground I am dazzled by the perpetual summer sunlight, but I know this is prime aurora territory, and if I were here in the winter darkness I would see the lights for real. This is the very place where Earth's magnetic field lines guide radioactive beams from space into the waiting arms of the ionosphere; where the aurora borealis shows the protective power of our air in action.

Clamped to the sea floor of our ocean of air, I try to picture its uppermost layers. But in spite of everything that I've read, I can scarcely believe that air too thin for me to breathe is yet strong enough to fend off everything that space can throw at us.

Yet it is. In October 2003, a series of explosions rocked the outer surface of the sun. A massive flare flash-fried Earth with x-rays equivalent to five thousand suns. A slingshot of plasma barrelled towards us at 2 million miles an hour. The radioactivity it contained was the equivalent, said one scientist, of taking every nuclear warhead that has ever been made – not exploded, mark you, but *made* – and detonating them all at once.

And yet nobody on Earth felt a thing (though perhaps you were lucky enough to see the light show). The most massive solar flare since records began and one of the biggest radioactive maelstroms in history together met a far more formidable foe. They each arrived, and then, one by one, they simply bounced off . . . thin air.

# NOTES

## *Chapter 1*

1 The intellectuals thus occupied claimed a Greek tradition for their methods, which would surely have surprised that arch-experimentalist Aristotle.
2 See Galileo's *Dialogues Concerning Two New Sciences.* The 'two new sciences', by the way, were the resistance of solid bodies to breaking, and a treatment of all forms of motion.
3 Contrary to legend, he didn't actually do this from the tower of Pisa.
4 He was too high by a factor of two, but it was still surprisingly close.
5 This idea originated with Aristotle in the fourth century BC and had persisted ever since. For once in his life Galileo was trapped in the wrong by sticking to received wisdom instead of thinking for himself.
6 Though it's still not clear exactly who performed the experiment that was to become famous, Torricelli probably asked his close friend and fellow-disciple of Galileo, Vincenzio Viviani, to commission the apparatus and do the actual deed.
7 Opinion is still divided on who came up with the idea of using mercury – it could have originated with Torricelli, or come from Viviani or even Galileo himself. In one copy of his *Dialogues,* after the part about the limited height to which water can be drawn up by a suction pump, Galileo seems to have dictated some margin notes to Viviani to the effect that other liquids should show a similar effect but to a lesser or greater height depending on the relative weights of the liquids, and he specifically mentioned wine, oil and quicksilver. See Middleton, *The History of the Barometer,* p. 20.

8 Middleton, *The History of the Barometer*, p. 24.

9 Another of Boyle's pieces, 'Upon the Eating of Oysters', describes how two fictitious friends discuss the unfairness of regarding other nations' customs as barbaric and yet failing to see how outsiders might view one's own habits: 'We impute it for a barbarous custom to many nations of the Indians,' says one, 'that like beasts they eat raw flesh. And pray how much is that worse than our eating raw fish, as we do in eating these . . . oysters whole, guts, excrement and all?' In reply, his companion says, 'you put me in mind of a fancy of your friend, Mr Boyle,' and goes on to describe Boyle's desire to write a short romantic story set in some South Pacific Island, governed by rational utopian laws, and how a native of those islands could travel throughout Europe and return home with bemused descriptions of our own strange, extravagant customs. Though Swift never acknowledged the debt, this could well have been the image that inspired *Gulliver's Travels*. Quoted, for example, in More, *The Life and Works of the Honourable Robert Boyle*, p. 158.

10 Pilkington, *Robert Boyle, Father of Chemistry*, p. 138.

11 This is about 3,750 million million tonnes, compared to the 5,600 million million tons quoted in Lyall Watson, *Heaven's Breath* (London: Hodder & Stoughton, 1984), p. 22. In 1648, Pascal had also persuaded his brother-in-law to take the quicksilver bath and glass tube up and down a mountain, to show that air weighs less at higher altitudes. Pascal's brother-in-law managed this with great success, though it can't have been easy to juggle the awkward contraption of bath, tube, quicksilver and all at the top of the 4,800-foot Puy-de-Dôme, where he repeated the experiment five times, 'once under cover in the little chapel which is there, or out-of-doors, once in a sheltered place, once in the wind, once while the weather was fine, and once during the fog and rain which came up now and then,' (Middleton, *The History of the Barometer*, p. 51).

12 Boyle suffered from ill health for much of his adult life, but he responded to it in typically practical fashion. To protect himself from chills he had commissioned a range of cloaks to suit every possible variation in the weather, and before he went out he decided which one to wear by consulting a recently invented instrument known as a thermometer.

13 Actually this wasn't quite what Boyle had expected. Torricelli had reported that the mercury remained some 26 or 27 inches above the surface, rather than 29½. If the mercury was being held up by the

downward weight of the atmosphere, why should there be such a difference? After all, exactly the same blanket of air was pressing down on both England and Italy. Could there be some problem with the apparatus, or, worse still, the theory? Before Boyle could toy with too many alarming ideas along these lines, he discovered that the answer was more to do with lack of intra-European co-ordination than any misbehaviour of the atmosphere. 'Our English Inches,' he noted with relief, 'are somewhat inferior in length to the digits made use of in Foreign Parts.'

14 The mercury never quite dropped to the level in the box because the pump couldn't quite succeed in extracting all the air. There was always a leak here or there, no matter how clever Hooke's design. But it fell quite far enough to satisfy Boyle, and ultimately the rest of the world.

15 He did however have an unfortunate tendency to overdose his descriptions. Why, after all, use one word when thirty would do? You could imagine him, dictating to his hapless secretaries, his sight too clouded to write for himself but his mind racing with ideas, determined to leave no room for error or doubt, and another thing, and another thing, and I must remember to mention yet another. An individual sentence would often contain well over one hundred words. For instance, from the introduction to *New Experiments Physico-mechanical Touching the Spring of Air*, here is Boyle's explanation for why he decided to study air. (Check out the second sentence – it has 121 words!):

> I am not faintly induced to make choice of this subject by these two considerations: the one, that the air being so necessary to human life, that not only the generality of men, but most other creatures that breathe, cannot live many minutes without it, any considerable discovery of its nature seems likely to prove of moment to mankind. And the other is, that the ambient air being that, whereto both our own bodies, and most of the others we deal with here below, are almost perpetually contiguous, not only its alterations have a notable and manifest share in these obvious effects, that men have already been invited to ascribe thereunto (such as are the various distempers incident to human bodies, especially if crazy in the spring, the autumn, and also on most of the great and sudden changes of weather) but likewise the further discovery of the nature of air will probably discover to us, that it concurs more or less to the exhibiting of many

phaenomena, in which it hath hitherto scarce been suspected to have any interest. So that a true account of any experiment that is new concerning a thing, wherewith we have such constant and necessary intercourse, may not only prove of some advantage to human life, but gratify philosophers by promoting their speculations on a subject, which hath so much opportunity to solicit their curiosity.

16 Boyle also discovered the law that was famously named after him, which states that if you squeeze any amount of air into a smaller volume, the pressure goes up. Squeezing air into a higher pressure also affects the temperature at which water can boil. In the rarefied air at the top of Mount Everest, water boils at around 160 degrees Fahrenheit, which is why it's impossible to make a decent cup of tea there. The reverse is true as well – at higher pressures, water boils at much higher temperatures. This is the principle behind the pressure cooker, which was invented by one of Boyle's contemporaries, Denis Papin, in 1682. The gentlemen of the recently created Royal Society in London had supper made in one, which, they wrote afterwards, 'caused much mirth amongst us and exceedingly pleased all the company'. See *Robert Boyle's Experiments in Pneumatics*, edited by James Bryant Conant.

## Chapter 2

1 See Crowther, *Scientists of the Industrial Revolution*, p. 181.

2 Even in Priestley's time, his invention of new soda water attracted much attention, as well as a prestigious prize. In time it came to the ears of the Admiralty, who had long been trying to find a way to combat the scourge of scurvy. It was already well known that something in vegetables combated the disease; but sailors could go for months without obtaining fresh food and half a ship's crew could succumb to the bleeding gums, listlessness and ultimately death before land was sighted and a new supply of greens obtained. Because rotting vegetable matter produced the same 'fixed air' that appeared during fermenting, one doctor had suggested that the air fixed inside plants must be what prevented scurvy. When Priestley happened across his handy way of forcing fixed air into water, the Admiralty was eager to know more. He was even offered a place on one of Cook's voyages. Perhaps fortunately, this time at least, his dissenting views meant that he was pulled from the rostrum at the

last minute. Soda water is of no use in curing scurvy – the answer, as we now know, was vitamin C.

3 Priestley also didn't realise that someone else, a young Swedish apothecary named Carl Scheele, had performed this same experiment a few years earlier. Scheele was self-effacing, and neither published his results nor tried particularly to interpret them. (He wrote a letter to French chemist Antoine Lavoisier, who appears later in this chapter, about his findings, to which Lavoisier never replied.) Nonetheless, there are still those who maintain that Scheele was the 'true' discoverer of oxygen. For an excellent fictionalised version of this story, see the play *Oxygen* by Carl Djerassi and Roald Hoffman. A good quick summary of the dates and events, with references, appears at http://antoine.frostburg.edu/chem/senese/101/history/faq/discovery-of-oxygen.shtml.

4 This quote is also sometimes attributed to his favourite mistress, Madame de Pompadour.

5 Aykroyd, *Three Philosophers*, p. 13.

6 Ibid., p. 89.

7 Ibid., p. 63.

8 Ibid., p. 106 and following.

9 Much of this first atmosphere was blasted away in a cosmic collision that created our moon, but it was quickly resupplied from within by gases pouring out of volcanoes.

10 When he read this section, chemist Andy Watson from the University of East Anglia became indignant on behalf of microbes, saying they weren't at all 'dull'. He has an excellent point, and I reproduce his comment in full: 'I'd like to point out that bacteria may not be very big or very fast, but in the field of biogeochemistry, their inventiveness makes animals look pretty dull. Bacteria exist that can use light, organic or inorganic chemical reactions as their energy source, that can use carbon dioxide, carbonate or organic carbon as their carbon source, that can live under aerobic or anaerobic conditions (or both), at temperatures from –1 degree to 400 degrees Celsius and pressures from 0 to at least 1,000 bars. And they do all the dirty jobs we'd rather not think about – without bacteria clearing up all the shit we animals produce, the earth as we know it would be finished in the blink of an eye!' Of course, Andy's right, though the point also remains that the world would look much less interesting if there were nothing but microbes on its surface.

11 One possible explanation is that evolution was kicked into action by mighty ice ages that blanketed the entire planet. See my book *Snowball Earth* (London: Bloomsbury, and New York: Crown, 2003).
12 See Lane, *Oxygen*, p. 125.
13 Ibid., p. 310. Charmingly, Lane says that he smokes but plans to give up the habit when he finishes writing his book.
14 Ibid. Note that there are many additional suggestions for why we need to have sex. One of the most persuasive is the notion that we are in a continual race to out-evolve parasites. See Matt Ridley's excellent book *The Red Queen* (London: Penguin, 1994) for more about this.
15 Aykroyd, *Three Philosophers*, p. 111.
16 Ibid., p. 111.
17 Crowther, *Scientists of the Industrial Revolution*, p. 231.
18 Ibid., p. 254.
19 About this time many other chemists had begun to notice nitrogen, though the actual 'discovery' of the element is usually attributed to a young Scottish chemist named Daniel Rutherford, who had isolated it a few years earlier.

## *Chapter 3*

1 Crowther, *Scientists of the Industrial Revolution*, p. 91.
2 Gibbon, author of *The Decline and Fall of the Roman Empire*, quoted in Crowther, p. 14.
3 Crowther, *Scientists of the Industrial Revolution*, p. 15.
4 Ramsay, *The Life and Letters of Joseph Black*, p. 22.
5 Donovan, *Philosophical Chemistry in the Scottish Enlightenment*, p. 194.
6 Allan and Schofield, *Stephen Hales, Scientist and Philanthropist*, p. 30.
7 Ibid., p. 43.
8 He was obliged to publish his thesis as a condition of receiving his degree. It's just as well, for given his reticence about publicising his own work we would otherwise know little about what proved to be a historic experiment.
9 Donovan, *Philosophical Chemistry in the Scottish Enlightenment*, p. 192.
10 Van Helmont was partly right in that some water had gone into providing the sap and in stiffening the new growth. But all of the solid matter had come instead from the air.

11 See, for instance, 'Life's a gas, if you're a plant', prepared by the United Kingdom's John Innes Centre for the Chelsea Flower Show. You can find this at http://www.jic.bbsrc.ac.uk/chelsea/handouts2004.htm.

12 See the excellent feature by Fred Pearce, 'The kingdoms of Gaia', in *New Scientist*, 16 June 2001, p. 30.

13 *New York Daily Tribune*, 23 October 1872, p. 6.

14 *John Tyndall: Essays on a Natural Philosopher*, p. 181.

15 Tyndall had a tendency to be oversensitive about criticism. At the early age of thirty-three he had been offered one of the two annual Royal medals by the Royal Society (which was a great honour, not only because the other recipient that year was Charles Darwin). He was about to accept when he heard that one member of the council had opposed the award and was complaining bitterly about it. Immediately he wrote a letter to the secretary of the society politely declining their honour. Huxley tried to convince him to change his mind, but Tyndall was obdurate. Later, Huxley wrote that at least it was a 'good sort of mistake', and added dryly that it was 'not likely to do harm by creating too many imitators'.

16 He was almost right. In fact the blue of the sky derives from scattering not from particles in the air but from the air molecules themselves, as Lord Rayleigh later proved.

17 See Weart, *The Discovery of Global Warming*, pp. 4–7.

18 Ibid., pp. 23–24.

19 We now know that somewhere between one-third and one-half of the $CO_2$ we've released has gradually disappeared into the ocean, which has considerably slowed down the build-up of $CO_2$ in the air.

20 The drop in carbon dioxide wasn't quite enough to cause the entire change in temperature, but this paper demonstrated once and for all that, added together with other greenhouse gases such as methane, it is a very important component.

21 See, for example, 'The "flickering switch" of late Pleistocene climate change', by K. C. Taylor et al., *Nature*, vol. 361 (4 February 1993), pp. 432–36.

22 D. A. Stainforth et al., *Nature*, vol. 433 (27 January 2005), pp. 403–406.

## Chapter 4

1 Watson, *Heaven's Breath*, p. 157.

2 See, for instance, 'Spora and Gaia: how microbes fly with their clouds', by W. D. Hamilton and T. M. Lenton, *Ethology, Ecology and Evolution*, vol. 10 (1998), pp. 1–16.

3 With due respect to the otherwise peerless Irving Berlin song 'They All Laughed', Columbus was not the first to realise that the world was round. In fact this had been known by all educated persons since the ancients.

4 When Columbus approached them, Isabella and Fernando were flush from their triumphant routing of the Moors at Granada. They had resolved to drive out all infidels from the Iberian peninsula, and one of Columbus's most persuasive arguments was that, with the supposed wealth he would bring from China, they could pursue their goals in a new crusade to wrest Jerusalem and the Holy Lands back from Moslem control. In the same crusading spirit, the monarchs had also banished any Spanish Jews who refused to convert to Christianity. A few tides before Columbus the last ship of refugees had set sail, bound for Moslem lands or for the Netherlands, the only Christian country prepared to receive them. Columbus would have been astonished to know that he was about to discover a continent that would eventually prove a refuge from this persecution.

5 Columbus's journal, pp. 9 and 13.

6 Not all the native peoples that Columbus encountered were quite so friendly, but that's another story.

7 Then, as now, this approach was a controversial one, and there are interesting parallels with the current popularity of 'Intelligent Design' as a supposed branch of science. One contemporary reviewer of *The Physical Geography of the Sea* wrote: 'It is now, we think, almost universally admitted, and certainly by men of the soundest faith that the Bible was not intended to teach us the truths of science. Our author, however, seems to think otherwise, and has taken the opposite side in the unfortunate controversy which still rages between the divine and the philosopher.' Another lauded Maury's 'strong and sincere religious feelings' but added: 'He unhappily does not see that in forcing Scripture to the interpretation of physical facts, he is mistaking the whole purport of the sacred Books, misappropriating their language and discrediting their evidence on matters of deep concern by applying it to objects

and cases of totally different nature.' See the introduction to *The Physical Geography of the Sea* by Matthew Fontaine Maury, 8th edition, edited by John Leighly (Cambridge, MA: Harvard University Press, 1963), p. xxvi.

8 Cox, *Storm Watchers*, p. 63.

9 Ibid., p. 63. Fortunately, Congress never approved Maury's request, and he disappeared from public view in 1861 at the start of the Civil War when he joined the Confederacy.

10 Ferrel's autobiography in *Biographical Memoirs of the National Academy of Sciences*, p. 296.

11 For an amusing description of some of the many unwitting disseminators of this myth, see http://www.ems.psu.edu/~fraser/Bad/BadCoriolis.html.

12 See three papers in *Bulletin of the American Meteorological Society*, vol. 47 (1966): Jordan, J. L., 'On Coriolis and the deflective force', pp. 401–403; Landsberg, H. E., 'Why indeed Coriolis', pp. 887–89; and Burstyn, Harold L., 'The deflecting force and Coriolis', pp. 890–91.

13 George Hadley himself got this far.

14 See Abbé's obituary in *Bulletin of the Philosophical Society of Washington*.

15 *Biographical Memoirs*, p. 298.

16 See the obituary by Professor W. M. Davis, *American Meteorological Journal*, vol. viii (1891), no. 8, p. 359.

17 Note that though ocean currents transport some of the heat necessary to alleviate this imbalance, air does most of the work. Oceans are responsible for about one-third and air for about two-thirds. See Barry and Chorley, *Atmosphere, Weather and Climate*.

18 Ferrel was the first person to realise why hurricanes never happen at the equator: it's the one place on Earth where the Coriolis force doesn't operate. Air has neither the urge to turn right nor left, so it can simply tumble into local low-pressure holes without whipping itself up into a hurricane frenzy.

19 A hurricane is typically about 400 miles across, compared with 900–1,800 miles for a mid-latitude storm. Also, though hurricanes tend to blow themselves out in a few days, mid-latitude weather fronts can last a week or more. See Barry and Chorley, *Atmosphere, Weather and Climate*.

20 'New England Weather', 1876, quoted in Watson, *Heaven's Breath*, p. 45.

21 Sterling and Sterling, *Forgotten Eagle*, p. 154.

22 Ibid., p. 139.
23 Ibid., p. 6.
24 Ibid., p. 153.
25 Ibid., p. 158.
26 Air has 0.035 per cent of Earth's water, which is $1.3 \times 10^{18}$ cubic metres, enough to coat Earth in a mere 2.5 centimetres of rain.

## *Chapter 5*

1 W. N. Hartley, 'On the absorption of solar rays by atmospheric oxygen', *Journal of the Chemical Society*, vol. xxxix (1881), p. 111.
2 When Midgley arrived it was the Dayton Engineering Laboratories Company (Delco), but it changed to General Motors four years later.
3 Haynes, *Great Chemists*, p. 1592.
4 See *Dictionary of Scientific Biography*.
5 The first commercial refrigeration system was patented in 1873, but they had only recently begun to be manufactured on an industrial scale.
6 Haynes, *Great Chemists*, p. 1595.
7 Ibid., p. 1596.
8 Cited in *Biographical Memoir of the National Academy of Sciences*, vol. xxiv, no. 11 (1947), pp. 361–80, by Charles F. Kettering.
9 Here I've paraphrased McNeill, who, in *Something New Under the Sun*, said that Midgley had 'more impact on the atmosphere than any other single organism in earth history'.
10 Kettering, p. 375.
11 'Father Earth', by Michael Bond, in *New Scientist* (9 September 2000), p. 44.
12 Lovelock, *Homage to Gaia*, p. 133.
13 Ibid., p. 241.
14 Ibid.
15 Ibid., p. 191.
16 Ibid., p. 199.
17 J. E. Lovelock, R. J. Maggi and R. J. Wade, 'Halogenated Hydrocarbons In and Over the Atlantic', *Nature*, vol. 241 (19 January 1973), p. 195.
18 The actual reactions are much more complicated, and involve several intermediaries. See, for instance, the detailed description in Richard Wayne's brilliant textbook *Chemistry of Atmospheres*, 3rd edition (London: Oxford University Press, 2000).

19 I have borrowed this memorable image from Sharon Roan in *Ozone Crisis.*

20 Roan, *Ozone Crisis*, p. 2.

21 Ibid., p. 31.

22 Lovelock, *Homage to Gaia*, p. 205.

23 *Nature*, vol. 249 (28 June 1974), pp. 810–12. Lovelock's stratospheric measurements swiftly followed in the same journal.

24 A message that seems to have made its mark down the years. In 2003, I played this song to a group of Princeton juniors and seniors during a class I was teaching on environmental science writing and told them I'd give extra credit to anyone who could name the singer and song. In unison, they all chanted, 'Joni Mitchell: "Big Yellow Taxi."'

25 Roan, *Ozone Crisis*, p. 37.

26 'Father Earth' by Michael Bond, in *New Scientist*, (9 September 2000), p. 44.

27 Roan, *Ozone Crisis*, p. 4.

28 Rowland and Molina, 'Ozone Depletion 10 years after the alarm', *Chemical and Engineering News*, vol. 72, no. 33 (15 August 1994), pp. 8–13.

29 Roan, *Ozone Crisis*, p. 81.

30 Ibid., p. 124. Note: there were still some international rumblings. The infant United Nations Environment Program held the Vienna Convention in March 1985. It was modest in ambition, with only twenty signatories and no regulatory powers – nothing compared with what would follow the discovery of the ozone hole itself.

31 Four successive stations have been crushed by the snow, and the fifth one, though balanced on steel stilts, will soon have to be replaced.

32 It was not until 1997 that women were allowed to winter there. The first woman visitor was back in 1973, but she doesn't really count. She was the wife of a ship's captain, who stepped on to the ice after the officers' dinner to be photographed next to the penguins in her evening gown.

33 Roan, *Ozone Crisis*, p. 127.

34 J. C. Farman, B. G. Gardiner and J. D. Shanklin, 'Large losses of total ozone in Antarctica reveal seasonal ClOx/NOx interaction', *Nature*, vol. 315 (1985), pp. 207–10.

35 Heath later claimed his group had already spotted the spurious data by the time Farman's paper was published, and had been secretly

trying to interpret them. In any case, he had certainly missed his chance at one of the scientific scoops of the century.

**36** She told me this in an interview in London in mid-September 2001, when she had refused to be scared into cancelling her plans to fly across the Atlantic to London. She was on one of the first planes that flew after 9/11.

**37** Rowland and Molina, 'Ozone Depletion 10 years after the alarm', *Chemical and Engineering News.*

**38** They shared this prize with Paul Crutzen, who had first realised that the ozone layer might be vulnerable by working out that nitrogen oxides could also destroy ozone.

**39** Lovelock, *Homage to Gaia*, p. 391.

## *Chapter 6*

**1** Degna Marconi, *My Father, Marconi*, p. 8.

**2** Ibid., pp. 11–12.

**3** Ibid., p. 14.

**4** Ultraviolet and infrared are other examples of electromagnetic waves; see Chapter 5. But whereas they have wavelengths of tiny fractions of an inch, wireless waves can have a distance between successive peaks and troughs of several miles. Since the shorter the wavelength, the higher the energy, radio waves are the least energetic of all, which is why we can live in a world that is permanently crisscrossed with radio waves yet suffer no ill effects.

**5** Degna Marconi, *My Father, Marconi*, p. 29.

**6** Dunlap, *Marconi: The Man and His Wireless*, p. 80.

**7** Ibid., p. 78.

**8** Weightman, *Signor Marconi's Magic Box*, p. 40.

**9** Ibid., p. 41.

**10** Karl Baarslag, in *SOS to the Rescue*, quoted in earlyradiohistory.us/sec005.htm.

**11** Degna Marconi, *My Father, Marconi*, p. 94.

**12** Ibid., p. 105.

**13** Some people still doubt that Marconi heard his signal on that day. However, see *The Friendly Ionosphere*, by Crawford MacKeand (Montchanin, Delaware: Tyndar Press, 2001). MacKeand went to elaborate technical lengths to model the equipment that Marconi used, and concluded that it is highly feasible that he was right.

**14** Dunlap, *Marconi: The Man and His Wireless*, p. 99.

**15** Degna Marconi, *My Father, Marconi*, p. 104.

16 Ibid., p. 105.
17 Dunlap, *Marconi: The Man and His Wireless*, p. 107.
18 13 January 1902.
19 Dunlap, *Marconi: The Man and His Wireless*, p. 113.
20 Ibid., p. 117.
21 Ibid., p. 118.
22 He was 5 feet 4½ inches.
23 Nahin, *Oliver Heaviside, Sage in Solitude*, p. 17.
24 Ibid., p. 99.
25 Ibid., p. 168.
26 Searle, in *The Heaviside Centenary Volume*, p. 9.
27 Sir Edward Appleton, in *The Heaviside Centenary Volume*, p. 3.
28 Nahin, *Oliver Heaviside, Sage in Solitude*, p. 293.
29 J. A. Crowther, quoted by Searle in *The Heaviside Centenary Volume*, pp. 8–9.
30 Searle, in *The Heaviside Centenary Volume*, p. 94.
31 Searle, in *The Heaviside Centenary Volume*, p. 9.
32 About the same time, an American scientist, Arthur Kennelly, made a similar suggestion. Heaviside had earlier written a more detailed treatment of the idea and submitted it as an academic article to be published in the *Electrician*, but the article was never published. That could be why researchers at the time used the term 'Heaviside' layer or, at most, appended Kennelly's name to Heaviside's. See Ratcliffe's *Sun, Earth and Radio.*
33 Searle, in *The Heaviside Centenary Volume*, p. 8.
34 Sir George Lee, in *The Heaviside Centenary Volume*, p. 16.
35 Nahin, *Oliver Heaviside, Sage in Solitude*, p. 292.
36 Ibid., p. 292.
37 One contemporary commentator joked that, rather than facing years of isolation, any modern Robinson Crusoe would simply need to fire up his ship's wireless apparatus, 'call up the nearest stations and ships, and pass the time waiting for relief in listening to the latest stock exchange quotations'. See Francis A. Collins, *The Wireless Man, His Work and Adventures on Land and Sea* (New York: The Century Company, 1912), p. 118. A few years earlier, two ships had collided, and one of them, the *Republic*, eventually sank. Help arrived thanks to a wireless distress signal, but all passengers had already been transferred to the remaining ship, the *Florida*, which still managed to limp into port.
38 See Harrison's *The Story of the Ionosphere.*
39 Weightman, *Signor Marconi's Magic Box*, p. 230; this was the

modern equivalent of nearly sixty dollars for the first ten words and nearly four dollars a word thereafter.

40 'SOS' is a simple (dot dot dot/dash dash dash/dot dot dot), whereas 'CQD' is the more complex (dash dot dash dot/dash dash dot dash/dash dot dot).

41 Francis A. Collins, *The Wireless Man* (New York: The Century Company, 1912) p. 14.

42 *New York Times*, 28 April 1912.

43 Dunlap, *Marconi: The Man and His Wireless*, p. 188.

44 One of the messages in Bride's pile was from Jack Thayer's mother. His father had disappeared with the *Titanic*. Mrs Thayer's message had read: 'Let anyone meet us but not children. Hope gone.' The message was never sent. See Booth and Coughlan, *Titanic: Signals of Disaster*.

45 Degna Marconi, *My Father, Marconi*, p. 190.

46 Bride gained $1,000 and Cottam, $750. Each was earning about $350 per year; see Weightman, p. 257.

47 Clark, *Sir Edward Appleton*, p. 9.

48 At this early stage in his career, there was no college silver for Appleton on the breakfast table. Nor was there fancy accommodation. During the war, he had married a woman from Bradford, and when his new wife first saw the unprepossessing terraced house that he had rented for them in Cambridge, she burst into tears.

49 Clark, *Sir Edward Appleton*, p. 21.

50 Ratcliffe, *Biographical Memoirs*, p. 7.

51 Clark, *Sir Edward Appleton*, p. 34.

52 This was at a presidential address to the British Association in the 1950s. See Clark, *Sir Edward Appleton*, p. 45.

53 Ratcliffe, *Biographical Memoirs*, p. 9.

54 Collins, *The Wireless Man*, p. 100.

55 This was only a few months before Oliver Heaviside's death.

56 Appleton, quoted in Clark, *Sir Edward Appleton*, p. 54.

57 This crackling region of air also turned out to be more complicated than anyone had realised. Appleton had named Heaviside's invention the E-Layer, to stand for electricity. But he later found another one higher in the sky, which he called F and most other people called the Appleton Layer, and then another, flimsier, one that lay below the E-Layer, and naturally came to be known as D. (Appleton explained: 'I didn't use the letters A, B, or C because I felt I must leave a letter or two in case someone discovered other layers below the D-Layer. They haven't done so, so now it looks a

bit off to start with the D-Layer. However I admit it's my fault.') See Clark, *Sir Edward Appleton*, pp. 60–61.

58 During his lecture at the ceremony in Sweden he told an elegant little joke to amuse the assembled guests. They should not, he said, place too much faith in the scientific method. For there was once a scientist who fed his friends a drink made up of whisky and soda water, and carefully observed the results. The next night he fed the same friends rum and soda water, and the next night gin and soda water. On each occasion, the friends got drunk. The scientist concluded that the agent responsible for this must have been the one thing the drinks had in common: soda water. The story went down well. Appleton didn't discover until later that the Crown Prince, later King Gustavus VI Adolphus, who was seated next to Lady Appleton, drank nothing but soda water.

59 Clark, *Sir Edward Appleton*, p. 45.

## *Chapter 7*

1 Green and Lomask, *Vanguard: A History*, chapter 11, p. 8 (Web version).

2 Birkeland was amazingly prescient – he would also try to raise funds to work on methods to exploit atomic energy before most of the world had even dreamed of this. Modern atomic theory didn't yet exist, and few people realised that atoms could be subdivided. But in 1905, Einstein published his famous paper on special relativity showing that mass was just another form of energy, and Birkeland made the connection that would later give rise to both the nuclear power station and the nuclear bomb. In 1906 he wrote to a Swedish banker: 'The problem I propose to solve is to find a practical way to utilise atomic energy. Our most important energy sources are stored in the molecules. If we solve this problem, we can get more energy out of one kilogram of matter than we get out of 10,000 kg of coal today.' See Egeland and Burke, *The First Space Scientist*, p. 127. Birkeland said he realised the problem was difficult and admitted it might not work, but added: 'I have never had such a mind to take up a thing as I have with this problem.' Unfortunately, in this case he didn't get his money. The banker called his idea 'titanic' and 'alluring', but said that he needed to wait for Birkeland's other inventions to turn a profit. See, for example, Devik, 'Kristian Birkeland as I knew him', p. 5.

3 Egeland, *The Man and the Scientist*, p. 14.
4 Devik, 'Kristian Birkeland as I knew him', p. 3.
5 Devik poetically wrote that Birkeland became 'all fire and flame'.
6 Birkeland's furnace was eventually supplanted by the Haber process, which is the modern basis for nitrogen fertilisers and involves splitting nitrogen by means of an iron catalyst. But for several decades, his sparking furnace reigned supreme.
7 Brekke and Egeland, *The Northern Lights*, p. 111.
8 There have been persistent reports that auroras are accompanied on rare occasions by a hissing sound. Though this has been rubbished by scientists for years, recent research suggests there may be something in them. See Harriet Williams, 'Sizzling Skies', *New Scientist*, 6 January 2001, p. 14.
9 'Carrington's Flare', at www.istp.gsfc.nasa.gov/Education/whcarr.html.
10 See, for instance, Jago, *The Northern Lights*, p. 23.
11 Egeland and Burke, *Kristian Birkeland: The First Space Scientist*, p. 134.
12 Jago, *The Northern Lights*, p. 172.
13 Ibid., p. 199.
14 Ibid., p. 118.
15 He had also decided to try to measure the heights of the auroras by recording them from two adjacent mountaintops connected by telephone lines. He had hoped that if each team took a picture of the same aurora at exactly the same time, but from a slightly different position, simple geometry should tell him how high the lights were. However, his cameras didn't work properly, so the idea came to nothing.
16 Birkeland, *The Norwegian Aurora Polaris Expedition 1902–1903*, p. 3.
17 Note: in spite of his handicap, he did, however, become a world-famous oceanographer.
18 Birkeland, *The Norwegian Aurora Polaris Expedition*, p. 5.
19 Ibid., p. 6.
20 See Egeland and Burke, *Kristian Birkeland: The First Space Scientist.*
21 Ibid. p. 9.
22 Birkeland, *The Norwegian Aurora Polaris Expedition*, section 1, preface, and section 2, p. 608.
23 See, for instance, Jago, *The Northern Lights*, p. 81.
24 Ibid., p. 267.
25 Martin Walt in 'From Nuclear Physics to Space Physics by Way of

High Altitude Nuclear Tests', p. 255, in 'Discovery of the Magnetosphere', *History of Geophysics.*

**26** 'Energetic particles in the earth's external magnetic field', by James van Allen, p. 235, in 'Discovery of the Magnetosphere', *History of Geophysics.*

**27** Protons do this, too, but the auroras they make are not visible to the naked eye.

**28** The system even works for astronauts. Missions such as the space shuttle don't fly as high as you might think, and apart from the Apollo missions, every human space flight has taken place beneath the shelter of our atmosphere's outermost protective layer. Between *Apollo 16* and *17* came a solar outburst massive enough to give any lunar astronauts a lethal dose of radiation within ten hours. Fortunately, no one happened to be flying. Any future human space flight back to the moon, to Mars, or anywhere else outside the air's aegis would have to be very heavily shielded.

**29** The inner Van Allen belt is populated with protons that come from cosmic rays, rather than from the sun.

**30** He was slightly wrong about these – though there are indeed vertical and horizontal currents in the atmosphere, the ones whose influence he measured were only off-shoots of the real things. But nobody has felt the need to quibble, and the name stands.

**31** Christine Hallas, 'The James van Allen Papers'.

# SUGGESTIONS FOR FURTHER READING

## *Prologue*

The main information for Kittinger's spectacular leap comes from his own memoirs, in Joseph W. Kittinger, Jr., 'The Long, Lonely Flight', *National Geographic* (February 1985), pp. 270–76, and Joseph W. Kittinger, Jr., 'The Long, Lonely Leap', *National Georgraphic* (December 1960), pp. 854–73; as well as Johnny Acton's entertainingly written *The Man Who Touched the Sky* (London: Sceptre, 2002), and Craig Ryan's graphic and detailed *Pre-Astronauts: Manned Ballooning on the Threshold of Space* (Annapolis: Naval Institute Press, 1995).

'Man's Farthest Aloft', by Captain Albert W. Stevens in *National Geographic Society Stratosphere Series* vol. 2 (1936), pp. 173–216, gives a charming description of the state of the art before Kittinger.

## *Chapter 1*

The Institute and Museum of the History of Science in Florence has an excellent website called 'Horror Vacui', which gives thumbnail sketches of the main characters in the discovery of the weight of air, and also has some nice pictures. You can find it at www.imss.fi.it/vuoto/.

Though there are hundreds of books about Galileo, most of them focus on his earlier life and his more famous discoveries, and few mention his experiments on air. The best way to find out what he did is to read his own highly entertaining words. The experiments described here are in his book *Dialogues Concerning Two New Sciences*, first published in Leiden in 1638. The version I used was translated by

H. Crew and A. de Salvio (New York: Macmillan, 1914). Nearly four centuries after the book was written, it still makes a great read.

One of the best sources for the relationship between Torricelli and Galileo, and for much else about the early development of pneumatics, is W. E. Knowles Middleton's *The History of the Barometer* (Baltimore: Johns Hopkins Press, 1964). Though – unlike Galileo's books – it isn't written in the world's most entertaining style, the book is comprehensive and clear, and has some gorgeous facsimiles of the original letters and diagrams. There is also a useful entry on Torricelli in the *Dictionary of Scientific Biography*, editor-in-chief Charles Coulston Gillispie (New York: Scribner, 1970–80). Blaise Pascal's *Physical Treatises* is also a good source for Torricelli's work, as well as Pascal's own. I used the version translated by I.H.B. and A.G.H. Spiers (New York: Columbia University Press, 1937).

For Robert Boyle, a good starting point is Michael Hunter's excellent website on Boyle at www.bkk.ac.uk/Boyle. Hunter is an eminent Boyle scholar, and as well as good material on Boyle, his site is full of useful references. Many of the books written about Boyle over the years hover between dull and sycophantic, but a few are highly readable. Three of the best sources that I found are Roger Pilkington, *Robert Boyle, Father of Chemistry* (London: John Murray 1959), which is vivid but sensible; Louis Trenchard More, *The Life and Works of the Honourable Robert Boyle* (London: Oxford University Press, 1944); and R. E.W. Maddison, *The Life of the Honourable Robert Boyle* (London: Taylor & Francis, 1969), which contains plenty of good detail and is well referenced. Also good is Thomas Farrington's *A Life of the Honourable Robert Boyle FRS, Scientist and Philanthropist* (Cork, Ireland: Guy & Co. Ltd., 1917).

But as with Galileo, the best way to read Boyle is in his own (admittedly sometimes lengthy) words. Try *Robert Boyle's Experiments in Pneumatics*, edited by James Bryant Conant, Harvard Case Histories in Experimental Sciences (Cambridge, MA: Harvard University Press, 1967), which contains plenty of quotes but also good context and analysis. Also recommended is *Robert Boyle by Himself and His Friends*, edited by Michael Hunter (London: Pickering & Chatto Ltd, 1994), which contains Boyle's own biographical sketch of his early life as well as biographical comments from various friends, and even a rather fulsome address delivered at his funeral.

Best of all is Boyle's own masterwork, *New Experiments Physico-mechanical Touching the Spring of the Air and its Effects (Made for the Most Part in a New Pneumatical Engine)*, which contains all the wonderful details of his experiments with the air pump. This is where you will find

Boyle's descriptions of his proof that air has spring, as well as his experiments with honey bees and mice, and the ones with birds that so distressed his lady visitor.

## *Chapter 2*

In spite of the burning of his house and loss of his papers, enough of Joseph Priestley's considerable output survived to provide rich resources for those interested in finding out more about his life and work. Good places to start are 'Joseph Priestley' in *Great Chemists*, edited by Eduard Faerber (New York: Interscience Publishers, 1961), pp. 241–51; and Robert E. Schofield's *The Enlightened Joseph Priestley* (University Park: Pennsylvania State University Press, 2004).

Two excellent articles are 'Joseph Priestley: Public Intellectual', by Robert Anderson, in *Chemical Heritage Newsmagazine*, vol. 23, no. 1 (Spring 2005), and 'Priestley, the furious free-thinker of the enlightenment, and Scheele, the taciturn apothecary of Uppsala', by John W. Severinghaus, in *Acta Anaesthesiologica Scandinavica*, vol. 46, pp. 2–9 (2002).

Of Priestley's own extensive writing, I would recommend Joseph Priestley, *Autobiography of Joseph Priestley, Memoirs Written by Himself, an Account of Further Discoveries in Air* (Bath, England: Adams & Dart, 1970), and Joseph Priestley, *A Scientific Autobiography*, edited with commentary by Robert E. Schofield (Cambridge, MA: MIT Press, 1966).

W. R. Aykroyd, *Three Philosophers, Lavoisier, Priestley and Cavendish* (London: William Heinemann, 1935) is full of rich, lively descriptions and interesting insights. Another fascinating work is James Gerald Crowther, *Scientists of the Industrial Revolution: Joseph Black, James Watt, Joseph Priestley, Henry Cavendish* (London: Cresset Press, 1962).

For the science of oxygen, look no further than Nick Lane's marvellously rich and detailed *Oxygen: The Molecule that Made the World* (London: Oxford University Press, 2002).

There is a useful entry on Antoine Lavoisier in the *Dictionary of Scientific Biography*. For more detail, try Jean-Pierre Poirier, *Lavoisier: Chemist, Biologist, Economist*, translated from the French by Rebecca Balinski (Philadelphia: University of Pennsylvania Press, 1996). A bit drier, but still good, is Arthur Donovan, *Antoine Lavoisier: Science, Administration and Revolution* (Cambridge, England: Cambridge University Press, 1993).

## *Chapter 3*

The life of Joseph Black is well described in A. L. Donovan, *Philosophical Chemistry in the Scottish Enlightenment: The Doctrines and Discoveries of William Cullen and Joseph Black* (Edinburgh: Edinburgh University Press, 1975), and in Sir William Ramsay, *The Life and Letters of Joseph Black, MD* (London: Constable & Co., 1918). Also see the excellent book by James Gerald Crowther, *Scientists of the Industrial Revolution: Joseph Black, James Watt, Joseph Priestley, Henry Cavendish* (London: Cresset Press, 1962). *Great Chemists*, edited by Eduard Faerber (New York: Interscience Publishers, 1961), has chapters on both Joseph Black and Svante Arrhenius.

For more about Stephen Hales, see D.G.C. Allan and R. E. Schofield, *Stephen Hales, Scientist and Philanthropist* (London: Scolar Press, 1980).

The richest source of information about John Tyndall is *John Tyndall: Essays on a Natural Philosopher*, edited by W. H. Brock, N. D. McMillan and R. C. Mollan (Dublin: Royal Dublin Society, 1981). This book of essays examines Tyndall's life and work from many perspectives, some technical, some from the point of view of his religious, philosophical and social values.

Many acres of trees have been chopped down to produce books about global warming, and some of the most worthily sacrificed were for Spencer R. Weart's excellent *The Discovery of Global Warming* (Cambridge, MA: Harvard University Press, 2003). Also see http://www.aip.org/history/climate/co2.htm, which gives a brief but accurate overview of the discovery of the greenhouse effect, with thumbnail sketches of the main characters.

## *Chapter 4*

A fascinating discourse on almost every aspect of wind can be found in Lyall Watson's *Heaven's Breath* (London: Hodder & Stoughton, 1984). Among the many books about Christopher Columbus, I found some of the most useful to be Washington Irving, *The Life and Voyages of Christopher Columbus*, vol. 1 (London: Cassell & Co. Ltd., 1827), which is austere and ponderous but full of interesting detail; Samuel Eliot Morison's *Christopher Columbus, Mariner* (London: Faber & Faber, 1956), which is much livelier and more entertaining, though occasionally a bit misleading (for instance, yes, Columbus had red hair, but by the time he embarked on his voyages it had already turned white); and David A. Thomas's *Christopher Columbus, Master of the Atlantic* (London: André Deutsch, 1991).

But there's nothing like reading the descriptions of the voyages in Columbus's own words. For this, go to *Christopher Columbus, the Journal of His First Voyage to America*, which is available in many editions. I used the one translated and with notes by Van Wyck Brooks (London: Jarrolds Publishers, 1925).

Shy William Ferrel left few of his own words behind, but there is a useful description of his life in the *Dictionary of Scientific Biography*. There's also an excellent entry on Ferrel in John D. Cox's fascinating *Storm Watchers* (Hoboken, New Jersey: John Wiley & Sons, 2002), pp. 65–74.

As for his friends' recollections, there is a collection of memorial articles in the *American Meteorological Journal*, December 1891, vol. viii, no. 8, pp. 337–69. In the same journal, February 1888, vol. iv, no. 10, pp. 441–49, there is an obituary by Ferrel's friend Alexander McAdie. Another of Ferrel's closest friends, Cleveland Abbé, wrote an obituary in the *Bulletin of the Philosophical Society of Washington*, vol. 12, 1892, pp. 448–60. Most valuable of all, from the diffident Ferrel, is the brief outline of his life written himself after much urging from McAdie. This is in *Biographical Memoirs of the National Academy of Sciences*, vol. 3 (1895), pp. 265–309. The same reference contains a memoir by Cleveland Abbé and a list of Ferrel's publications. See also 'William Ferrel and American Science in the Centennial Years', by Harold L. Burstyn, in *Transformation and Tradition in the Sciences, Essays in Honor of I. Bernard Cohen*, edited by Everett Mendelsohn (Cambridge, England: Cambridge University Press, 1984), pp. 337–51.

And of course there is Ferrel's own essay, 'An essay on the winds and currents of the ocean', which is the first entry in 'Popular Essays on the Movements of the Atmosphere by Professor William Ferrel', published as number XII of the *Professional Papers of the Signal Service* (Washington DC, 1882).

For more about the science of the wind, see Roger G. Barry and Richard J. Chorley, *Atmosphere, Weather and Climate*, 8th edition (London and New York: Routledge, 2003). I would respectfully suggest that this is the best textbook ever written about the movements of the air and the way they affect weather; it's no wonder it's in its eighth edition and counting.

The best source on Wiley Post is Bryan B. Sterling and Frances N. Sterling's *Forgotten Eagle: Wiley Post, America's Heroic Aviation Pioneer*

(New York: Carroll & Graf, 2001). The book is now sadly out of print, but you can find it secondhand. Beware, though: a printing error meant that the first copy I bought was missing the crucial (and fascinating) description of Post's stratospheric flights. Fortunately, I found a sympathetic secondhand bookstore owner in Cape Cod, who patiently leafed through his copy to check that it was all there before sending it to me in London.

And for more about the disappearing plane, see BBC Horizon's *Vanished: The Plane that Disappeared*, which was broadcast on 2 November 2000. The transcript of this fascinating programme is available on the web at http://www.bbc.co.uk/science/horizon/2000/vanished.shtml.

## *Chapter 5*

Thomas Midgley's life and work are nicely described in an essay by William Haynes in *Great Chemists*, edited by Eduard Faerber (New York: Interscience Publishers, 1961), pp. 1589–97, as well as 'Thomas Midgley' in *Dictionary of American Biography, Supplement 3* (New York: Charles Scribner's Sons, 1941–45), pp. 521–23, and Charles Kettering's affectionate memoir of his friend in *Biographical Memoir of the National Academy of Sciences*, vol. xxiv, no. 11 (1947), pp. 361–80.

There is an excellent account of how it all went wrong and the rest of the ozone story in 'An environmental fairytale' by Aisling Irwin in *It Must Be Beautiful, Great Equations of Modern Science*, edited by Graham Farmelo (London: Granta Books, 2002). A longer, but engaging and very readable, account is in Sharon Roan's *Ozone Crisis: The 15-Year Evolution of a Sudden Global Emergency* (New York: Wiley, 1989). John McNeill's stern book *Something New Under the Sun: An Environmental History of the Twentieth Century* (New York: W.W. Norton & Co., 2000) contains a good section on 'Climate change and stratospheric ozone'. But the best read of all for the ozone wars and the rest of his extraordinary life is James Lovelock's marvellous autobiography, *Homage to Gaia: The Life of an Independent Scientist* (Oxford: Oxford University Press, 2000).

The Nobel prizes website is also a rich source of both technical and biographical information about the ozone laureates and the context in which they worked. See http://nobelprize.org/chemistry/laureates/1995.

## *Chapter 6*

Excellent descriptions of elves, jets and sprites can be found in two *New Scientist* feature articles: 'Bolts from the Blue' by Keay Davidson (19 August 1995, p. 32) and 'Rider on the Storm' by Harriet Williams (15 December 2001, p. 36). For more details on the bizarre science of the ionosphere, try J. A. Ratcliffe, *Sun, Earth and Radio, an Introduction to the Ionosphere and Magnetosphere* (London: Weidenfeld & Nicolson, 1970). J. A. Harrison's *The Story of the Ionosphere or Exploring with Wireless Waves* (London: Hulton Educational Publications, 1958) is fun for its relentlessly 1950s schoolboy tone. For those who want something much more serious and detailed, there is Robert W. Schunk and Andrew F. Nagy's *Ionospheres: Physics, Plasma Physics and Chemistry* (New York: Cambridge University Press, 2000).

Of the many books written about Marconi and his achievements, I would recommend W. P. Jolly's *Marconi* (London: Constable, 1972); Orrin E. Dunlap's *Marconi: The Man and His Wireless* (New York: The Macmillan Co., 1937); and, in particular, Gavin Weightman's engagingly written *Signor Marconi's Magic Box: How an Amateur Inventor Defied Scientists and Began the Radio Revolution* (London: HarperCollins, 2003). Degna Marconi, Marconi's daughter, also provides an interesting perspective in *My Father, Marconi* (London: F. Muller, 1962).

There are many fascinating memoirs written about the fabulous Oliver Heaviside. First stop would be the useful entry in the *Dictionary of Scientific Biography*. For a nice summary of Heaviside's scientific achievements, try the brief, polite obituary by A. Russell in *Nature*, vol. 115 (14 February 1925), pp. 237–38. More entertaining is *The Heaviside Centenary Volume* (London: Institution of Electrical Engineers, 1950), which contains a collection of articles on Heaviside's work and several commentaries on his personality. Look in particular for the affectionate description of Heaviside's oddities by his good friend G.F.C. Searle, who expanded on this in a full-length book crammed with personal reminiscences, *Oliver Heaviside, the Man* (Cambridge, England: CAM Publishing, 1988). Also excellent reading is Paul J. Nahin's *Oliver Heaviside, Sage in Solitude* (New York: IEEE Press, 1988).

So many books have been written about the *Titanic* disaster that I will highlight only a few. Try John Booth and Sean Coughlan's *Titanic: Signals of Disaster* (White Star Publications, 1993), and Walter Lord's *A Night To Remember* (London: Longmans Green & Co., 1958). Jack Thayer's memoir of the night, 'The Sinking of the SS *Titanic*', is vivid and shocking, but unfortunately also hard to find. It was originally published in 1940, and reprinted by 7C's Press in 1974, but both editions are now out of print.

For Appleton, by far the best source is Ronald W. Clark's *Sir Edward Appleton, G.B.E., K.C.B., F.R.S.* (Oxford: Pergamon, 1971). There are also useful entries in the *Dictionary of Scientific Biography* and in the memoir written by one of his students: J. A. Ratcliffe, *Biographical Memoirs of Fellows of the Royal Society*, vol. 12 (1966), pp. 1–21.

## *Chapter 7*

For the story of James van Allen's discovery of the Van Allen belts, an excellent place to start is Constance McLaughlin Green and Milton Lomask's *Vanguard: A History*, NASA SP-4202 (Washington DC: Smithsonian Institution Press, 1971). The book is now out of print, but it's available online at http://www.hq.nasa.gov/office/pao/History/SP-4202/cover.htm.

Van Allen's own account is in 'What Is a Space Scientist? An Autobiographical Example', in *Annual Review of Earth and Planetary Sciences*, June 1989. See also his article 'Radiation belts around the Earth' in *Scientific American*, vol. 200, no. 3 (March 1959), pp. 39–48, and his *Origins of Magnetospheric Physics* (Washington DC: Smithsonian Institution Press, 1983). Most of the information on Van Allen in this chapter came from these four sources.

More technical is the weighty 'Magnetospheric Currents', in *Geophysical Monograph 28*, edited by Thomas A. Potemra (Washington DC: AGU, 1983).

A good collection of articles can be found in 'Discovery of the Magnetosphere', edited by C. Stewart Gillmor and John R. Spreiter, in *History of Geophysics*, vol. 7 (Washington DC: AGU, 1997). These are partly technical, partly biographical, and include an entry by Van Allen himself. 'The James van Allen Papers' by Christine Halas is at http://www.lib.uiowa.edu/spec-coll/Bai/halas.htm. This discusses the University of Iowa Van Allen collection, and also contains some entertaining anecdotes about Van Allen himself.

For more information about the science of the magnetosphere, as well as further details on the history of its discovery, see David P. Stern's excellent article in *Reviews of Geophysics*, vol. 40, no. 3 (September 2002), pp. 1–30. This is also available on the web at http://www.phy6.org/Education/bh2_2.html. Also accessible and entertaining are 'Watch out, here comes the sun' by Hazel Muir, *New Scientist*, 3 February 1996, p. 22, for space weather, and 'Into the sphere of fire' by

Stephen Battersby, *New Scientist*, 2 August 2003, p. 30, for a marvellous description of the magnetosphere's bizarre habits.

Kristian Birkeland's own description of his aurora field trips is in *The Norwegian Aurora Polaris Expedition 1902–1903*, vol. 1, sections 1 and 2 (Oslo: H. Aschehoug & Co., 1913). The introduction gives rich details of the troubles faced by the teams. (The book is labelled 'Volume 1' because there should have been a second volume dealing explicitly with auroras. However, it was never published, and some have speculated that the manuscript went down with a ship carrying Birkeland's affairs after his death.)

For more on Birkeland's life, see A. Egeland and E. Leer, 'Professor Kr Birkeland: His life and work', in *IEEE Transactions on Plasma Science*, vol. PS-14, no. 6 (December 1986). This contains many biographical details, as well as fascinating information on Birkeland's solar system studies, such as the functioning of the sun and the rings of Saturn. Other useful papers by Alv Egeland include his excellent 'Kristian Birkeland: The Man and the Scientist', in 'Magnetospheric Currents', *AGU Geophysical Monograph 28* (Washington DC, 1984), pp. 1–16. This monograph also contains several other useful papers, notably A. J. Dessler's 'The evolution of arguments regarding the existence of field-aligned currents', pp. 22–28. For an entertaining account of the electromagnetic gun fiasco, see Egeland's paper 'Birkeland's Electromagnetic Gun: A Historical Review', *IEEE Transactions on Plasma Science*, vol. 17, no. 2 (April 1989), pp. 73–82. Egeland also co-wrote a diligently researched and documented biography of Birkeland with William J. Burke, *Kristian Birkeland: The First Space Scientist* (Dordrecht, Netherlands: Springer, 2005).

Lucy Jago has also written a biography of Birkeland: *The Northern Lights: How One Man Sacrificed Love, Happiness and Sanity to Unlock the Secrets of Space* (London: Hamish Hamilton, 2001). This book is fascinating, well written and clearly extensively researched. But beware: exasperatingly, the author has included no direct references or footnotes and says herself that she has 'telescoped' some events to help the story along and also made 'reasonable' assumptions in some unspecified cases. Unfortunately, this makes it hard to trust any of the details unless they're corroborated by another source.

For personal reminiscences of Birkeland's unique style, see the memoir written by his former lab assistant, Olaf Devik: 'Kristian Birkeland as I knew him', in *Birkeland Symposium on Aurora and Magnetic Storms*, edited by A. Egeland and J. Holtet (Paris: CNRS, 1968).

And finally, for more on the northern lights themselves, see Asgeir Brekke and Alv Egeland, *The Northern Lights*, translated by James Anderson (Oslo: Grøndahl Dreyer, 1994). This is a marvellous tapestry weaving together myths and legends, literature, history and science to give a full portrait of human responses to the northern lights. There are also some fabulous illustrations.

# ACKNOWLEDGEMENTS

I have been writing stories about parts of the atmosphere for over a decade, and yet it might never have occurred to me to look at the air as a whole if it hadn't been for the questions people kept asking me. Thanks, therefore, to Fred Barron, who wanted to understand about the winds, and to Simon Singh, who wondered how we first discovered that the atmosphere has layers, and what each layer really does. It was Simon, too, who suggested that I calculate the weight of the air in an 'empty' concert hall, the results of which surprised me much more than they did him. Thanks also to the many people who wanted to know more about how the atmosphere was changing. After ten years of preaching about climate change only, it appeared, to the converted, I'm thrilled that it has finally become such a hot topic.

However, it was the image of Joseph Kittinger's jump that finally convinced me to write about air. Thanks to Jonathan Renouf who alerted me to Kittinger's existence, and lent me tapes of a fabulous BBC programme about the atmosphere which contained some of the original shots from Kittinger's on-balloon camera as he fell. (Jonathan was also helpful with information about *Stardust*, the plane that vanished in the southern jet stream, since he was writer and producer of the *Horizon* programme on the subject.)

As I watched the video of Kittinger's extraordinary leap, I stared at the thin blue line of the atmosphere that hovered over

Earth's curved horizon. And after he jumped I watched him, floating on a sea of what he later said seemed like nothing. I was enchanted by the paradox of this atmosphere that Kittinger fell through. How can something so delicate also be so powerful? An unsung hero full of fragility and fire: what more could a writer ask for?

And so I set to work. And the more I investigated, the more I realised that an extraordinary cast of characters had collaborated to show us the power of air. One of my favourites, and the hardest to pin down, was William Ferrel, the diffident West Virginia farmer who figured out the trade winds by drawing circles with a pitch fork on his barn door. Thanks to John Cox for help with sources about Ferrel's life and work. Thanks also to the fine folks at the National Academy of Sciences, for sending me a copy of the sole autobiographical sketch that this remarkable man left behind when I had exhausted all other possible sources. (How could the British Library have two sets of the entire series in which this sketch appeared, and yet that volume be missing in each? And how could I fail to find it in the few remaining collections in the rest of the UK, and even at the few US universities I tried in desperation? I half-suspected the shy Ferrel of reaching beyond the grave to efface the few written details that remained of his life before I could read them.)

In all other respects, the British Library was as valuable as I have come to expect, as were the ever-helpful staff of Reading Room Science 2. Thanks also to the staff and curators of the wonderful London Library. For some reason this element of literary London remains largely hidden. And yet the charming, archaic building in St James's Square, London houses a treasure of ancient and modern volumes in its labyrinthine stacks.

Best of all, you can borrow books from the London Library and take them away with you. In my case I took them to France, where I wrote the first half of this book in the tiny but welcoming village of Condeissiat in Ain. Thanks especially to the Famille Sinardet, at Les Fausses: Hélène, Jean-Chris and

Hubert, not to mention Sammy, Choupette and Clochette. Hélène kept me supplied with her peerless tartes à la crème, the boys with cheese, wine and good English tea, while the animals provided me with distractions when I needed them, and left me alone (more or less) when I didn't. And Hubert greeted me at the end of each working day with the simple question 'How many words?', which focuses the mind tremendously.

The second half of the book I wrote back in London during what seemed like a very long, cold winter. (Though Hubert was now away in Antarctica, he helpfully left me with an iPod recording saying: ('How many words?' . . . pause . . . 'Well done!' The fact that he didn't bother to record an alternative in case the words hadn't come also proved unexpectedly encouraging.)

During that winter Fred Barron – the best of neighbours and of friends – made me laugh and fed me steaks and classic movies to keep up my strength. He has been fabulous throughout my writing of this book. From the beginning he has shared my excitement, first over the topic and then the characters. When I came across a new story he was usually the first to hear it, and I think the people I wrote about began to seem as real to him as they did to me. Being a comedy writer, he was also quick to spot the places where I'd unwittingly blown my own punchline.

David Bodanis, my friend and mentor, was also there from the start. Chapter 1 especially owes a great deal to his excellent advice. He also helped me immeasurably with what I at least find to be the hardest part of writing a book: the beginning.

People who read and commented on the manuscript include Robert Coontz, Richard Stone, John Vandecar, Karen Southwell, Dominick McIntyre, Fred Barron, David Bodanis, Elan McAllister, Michael Bender, Andy Watson, John Mitchell and David Rind. Rosa Malloy especially employed her considerable talents for spotting where explanations grew tangled or stories overlong. All of their comments and criticisms improved the manuscript substantially; of course any remaining errors are my own.

Thanks to my agent, Michael Carlisle, for his unstinting efforts on my, and air's, behalf. And thanks especially to two of the best editors in the business: Andrea Schulz at Harcourt and Bill Swainson at Bloomsbury. Together they helped me shape the manuscript where it needed shaping, yet refrained from fixing any parts that weren't broken. (Though I am fairly sure they weren't in collusion, their opinions also coincided uncannily, which was very reassuring.)

Many other people have lived graciously along with me, while I was living with Air. Thanks for particular support and tolerance go to John Vandecar, Karen Southwell, Stephen Battersby and Dominick McIntyre.

Finally, thanks to my wonderful family: Rosa, Helen, Ed, Christian, Sarah, Damian, Jayne and the kids, and Hubert. Only you know how little I could do without you.

# INDEX

## A NOTE ON THE AUTHOR

Gabrielle Walker has a PhD in chemistry from Cambridge University and has taught at both Cambridge and Princeton universities. She is a consultant to *New Scientist*, contribues frequently to BBC radio and writes for many newspapers and magazines. The author of *Snowball Earth* and writer and presenter of BBC Radio 4's *Planet Earth under Threat*, she lives in London and France.

## A NOTE ON THE TYPE

The text of this book is set in Adobe Garamond. It is one of several versions of Garamond based on the designs of Claude Garamond. It is thought that Garamond based his font on Bembo, cut in 1495 by Francesco Griffo in collaboration with the Italian printer Aldus Manutius. Garamond types were first used in books printed in Paris around 1532. Many of the present-day versions of this type are based on the *Typi Academiae* of Jean Jannon cut in Sedan in 1615.

Claude Garamond was born in Paris in 1480. He learned how to cut type from his father and by the age of fifteen he was able to fashion steel punches the size of a pica with great precision. At the age of sixty he was commissioned by King Francis I to design a Greek alphabet; for this he was given the honourable title of royal type founder. He died in 1561.